6

D1756014

THE SUBSIDY SCANDAL

How *Your* Government Wastes *Your* Money to Wreck *Your* Environment

Charlie Pye-Smith

Earthscan Publications Limited
London • Sterling, VA

First published in the UK and USA in 2002 by
Earthscan Publications Ltd

ISBN: 1 85383 902 7

Typesetting by JS Typesetting Ltd, Wellingborough, Northants
Printed and bound by Creative Print and Design (Wales), Ebbw Vale
Cover design by Danny Gillespie

For a full list of publications please contact:

Earthscan Publications Ltd
120 Pentonville Road
London, N1 9JN, UK
Tel: +44 (0)20 7278 0433
Fax: +44 (0)20 7278 1142
Email: earthinfo@earthscan.co.uk
http://www.earthscan.co.uk

333.73158/PYE

22883 Quicksilver Drive, Sterling, VA 20166-2012, USA

A catalogue record for this book is available from the British Library

Library of Congress Cataloging-in-Publication Data

Pye-Smith, Charlie, 1951–.
  The subsidy scandal: how governments squander public money and
destroy the environment / Charlie Pye-Smith.
      p. cm.
  Includes bibliographical references and index.
  ISBN 1-85383-902-7 (hardback)
    1. Subsidies–United States. 2. Subsidies–Environmental aspects–United
States. 3. Waste in government spending–United States. I. Title.
HC110.S9 P94 2002
338.973'02–dc21

                                                        2002004457

Earthscan is an editorially independent subsidiary of Kogan Page Ltd and
publishes in association with WWF-UK and the International Institute for
Environment and Development

This book is printed on elemental chlorine-free paper

CONTENTS

Acronyms and Abbreviations

ARC   Appalachian Regional Commission
BLM   Bureau of Land Management
BSE   bovine spongiform encephalitis
CAFO   concentrated animal-feed operation
CAP   Common Agricultural Policy
Cdn   Canadian
DFO   Department of Fisheries and Oceans (Canada)
DNR   Department of Natural Resources (US)
DoE   Department of Energy (US)
EAA   Everglades Agricultural Area
EPA   Environmental Protection Agency (US)
ESA   Endangered Species Act (US)
EU   European Union
FAO   Food and Agriculture Organization (United Nations)
FoE   Friends of the Earth
FPI   Fisheries Products International Limited
GAO   General Accounting Office (US)
IIED   International Institute for Environment and
        Development
MAP   Market Access Program

MIT	Massachusetts Institute of Technology
MOGA	Michigan Oil and Gas Association
NAFTA	North American Free Trade Agreement
NEPA	National Environmental Protection Act (US)
NMA	National Mining Association
OECD	Organisation for Economic Co-operation and Development
PAC	political action committee
PDA	percentage depletion allowance
PERC	Political Economy Research Center
PIRG	Public Interest Research Group (US)
ppbn	parts per billion
R&D	research and development
ROD	Record of Decision
SAFCA	Sacramento Flood Control Agency
SEACC	South-East Alaska Conservation Council
TAG	Technical Assistance Grant
USDA	US Department of Agriculture
WRI	World Resources Institute
WTO	World Trade Organization
WWF	World Wildlife Fund

ACKNOWLEDGMENTS

An enormous number of people helped and entertained me during my travels round the United States and Canada. Some asked not to be mentioned by name, but my thanks go to them as well as to all of the following.

In Washington, DC: Anna Aurilio and Lexi Shultz of the United States Public Interest Research Group; David Hirsch, Gawain Kripke, Lynn Erskine, Courtney Cuff, and Brent Blackwelder of Friends of the Earth; Jill Lancelot, Taxpayers for Common Sense; David Conrad, National Wildlife Federation; Dan Becker, Sierra Club; Robert Ferris and Caroline Kennedy of Defenders of Wildlife; Philip Voorhees and Neil Evans of the National Parks Conservation Association; Andrew Brecher, Concord Coalition; Celia Boddington, Bureau of Land Management; David Freestone, World Bank; Ric Fenton and Karen Batra, National Mining Association; Alan Septoff and Sue Brackett of the Mineral Policy Center; Congressman Dan Miller.

In Alaska: Matt Zencey, Alaska Rainforest Campaign; Diana Rhoades; The Boat Company and the crew of *M/V Observer*; Steve Behnke, Alaska Wilderness, Recreation and Tourism Association; Thomas Gemmell, United Fishermen of Alaska; Dale Kelley, Alaska Trollers Association; Amelia Jenkins, Association of Forest Service Employees for Environmental Ethics; Lana Shea Flanders, Alaska

Department of Fish and Game; Stan Filler, mayor of Sitka; Jim Clark; Mark Wheeler and Buck Lindekugel, South-East Alaska Conservation Council; Richard Nelson; Jack Lorrigan; Jim Caplan, USDA Forest Service; Brian McNitt and colleagues at the Sitka Conservation Society.

In Colorado: Ignacio Rodriguez; Camille Price; Wendy Mallot; Sara Jones; Bob Green; Jo Gallegos; Dan Randolph, Mineral Policy Center.

In West Virginia: Eddie Canterbury, Jim Kingsbury and Charlie Phillips of the Corridor H Action Committee; Hugh and Ruth Rogers and Bonni McKeown of Corridor H Alternatives; Chuck Merritt; Michael Kline.

In Montana: Aimee Boulanger, Mineral Policy Center; Tom Skeele and David Gaillard, Predator Alliance; Becky Weed, Growers Wool Cooperative; John Parr; Bob Gilbert, Montana Wool Growers Association; Beth Emter, Montana Stockgrowers Association; John Wilson, Montana Land Reliance; Bishop Grewell, Political Economy Research Center; David Jaynes and Sandy Brooks, Bureau of Land Management; Chum and Sally Howe, Ross Peak Ranch; Wally McRae, Rocker Six Cattle Co.; John Youngberg, Farm Bureau; George Wuerthner; Larry Handegard, Wildlife Services; Jim Hamilton; the organizers of the Montana Cowboy Poetry Festival.

In Michigan: Keith Schneider, Hans Voss, Patty Cantrell, and Kelly Thayer of the Michigan Land Use Institute; Kitty Myers; Frank Mortl, Michigan Oil and Gas Association; Greg Fogle, O.I.L. Energy Corps; Tom Edison; Ed Lennington; Donna Hardies; Mike Miller, Miller Energy Inc.; Anne Woiwode, Sierra Club; Bernadette Fletcher; Gerry and Louise Burns.

In New Mexico: Frank Dubois, New Mexico Department of Agriculture; Betty Hyatt, People for the USA; Leadrue Hyatt; Nick Ashcroft, New Mexico State University; Andrea Martinez, Gila National Forest; Jodie Savary; John Horning, Forest Guardians; Michael Sauber, Gila Watch; Sally Smith; Mike Head; Tricia London;

Roger Kline, Gila National Forest; Don Manning, United Steel-workers of America.

In Florida: Mark Kraus, National Audubon Society; Charles Lee, Florida Audubon Society; Gene Duncan, with the Miccosukee Tribe of Indians of Florida; Mary Barley; Robert Buker, Judy Sanchez, Roger Moss, Murray Brinson, and Don Griffin of the United States Sugar Corporation.

In California: William Mueller, Sacramento Metropolitan Chamber of Commerce; Bob Childs and Walter Yep of United States Army Corps of Engineers; Ron Stork and Charlie Casey of Friends of the River; Lowell Jarvis, Placer County Water Agency; Bruce Cosgrove, Auburn Area Chamber of Commerce; Gary Estes; Eric Peach; Richard Robinson, advisor to Congressman John Doolittle; Clyde MacDonald; Joanna Wald, Natural Resources Defense Council; Ed Tiederman; Zeke Grader, Pacific Coast Federation of Fishermen's Federations.

In Newfoundland: Ray Andrews; Tom Best; Barbar Neis, Bill Schrank, and Bill Montevecchi of Memorial University of Newfoundland; Reg Anstey, Fish, Food and Allied Workers; Art May; George Winters; Bernard Brown; John Efford; Cyril Ryan; Wallace Tucker; Owen Myers.

Jonathan Sinclair Wilson and Pascale Mettam of Earthscan have been endlessly supportive during the writing of this book, and were it not for the guidance and enthusiasm of Lloyd Timberlake, it would never have seen the light of day. I owe the greatest debt of gratitude to the AVINA Foundation, whose generosity made this all possible.

Charlie Pye-Smith
June 2002

INTRODUCTION: ON THE SUBSIDY TRAIL

Ignacio Rodriguez, a craggy-faced man in his mid-seventies, motioned down to the Alamosa River as we were leaving his small ranch in southern Colorado. "We used to catch trout and barbecue them down there," he said, pointing towards a narrow shelf of land with beaver-gnawed stumps along the water's edge. "But the gold mine killed the river. There's not been a living thing in it for seven years now."

As the morning sun rose over the creamy green oceans of sage brush, we headed west along a sunflower-lined dirt road. Soon we were driving through fine countryside, as dramatic as any you will see in the American West and rich in wildlife. Last night a mountain lion had set the ranch dogs barking, and soon after we began climbing through the aspen and spruce, we slowed to let a black bear amble across the road. Rodriguez said there was good hunting up here, so we kept our eyes skinned for elk and mule deer, but no one fished the Alamosa any more, and it was so acid only a fool would swim in it. Some two hours after we left his ranch we crested a flank of open land at around 11,000 feet, to be confronted by a scene of utter devastation. Half a mountainside had been gouged away.

"This," said Rodriguez, with an expression of contempt, "is the disaster called Summitville." Over a century ago bands of fortune seekers with picks and shovels had dug for gold here, and on and off since then gold had been mined. "But it was always underground mining," explained Rodriguez. "That is, till a Canadian company called Gallactic Resources appeared on the scene." In the late 1980s Gallactic Resources created an open-pit mine and extracted gold using a technique known as cyanide-heap leach mining. Crude but effective, this involved digging out the rock with vast dirt movers and drizzling cyanide onto it to remove precious metals.

"Then one day in 1989," recalled Rodriguez, "we had a big fish kill down in the valley. I thought it was whirling disease, which sometimes kills the trout. Now I know better. It must have been cyanide." The mining company had failed to prevent the chemical getting into the Alamosa. The river recovered for a while, but there was a major leak of cyanide in 1992. This time, it killed all life in the river, from Summitville right down to the San Luis Valley some 20 miles away. The mine owners, having made handsome profits from Summitville, eventually declared themselves bankrupt and ran, leaving the government to clear up the mess and the farmers downstream to rue their misfortune. Irrigation equipment that used to last 25 years is now eaten away by the acid river water in six or seven. The sheep that graze in the valley have abnormally high levels of copper in their livers, and no one knows for sure what long-term effects the pollution will have on crops, livestock, or wildlife.

But it is not just a few hundred mainly Hispanic farmers—the Alamosa irrigates 40,000 acres—who are suffering from this abandoned gold mine. The American taxpayer is also losing out on a grand scale. Rodriguez explained that the taxpayer is footing part of the clean-up bill, which is set to soar above US$150 million. However, this figure pales beside a far greater give-away to the mining industry, which comes in the form of a subsidy that dates back to the 19th century. This particular plot of land used to be public land managed by the Forest Service, but under an archaic law, formulated at a time when the US government was encouraging the opening up of the West, it was bought by private mining interests for US$5 an acre. This practice of "patenting" has led to some extraordinary windfalls for the mining industry, with companies paying a few thousand dollars to gain access to reserves worth billions.

What makes this even more outrageous, in the view of Rodriguez and many others, is that under the 1872 Mining Law these companies pay no royalties for the exploitation of what is—or was —a publicly owned resource. If hard-rock mining companies had been charged an 8 percent royalty on minerals extracted from public lands—this is what the coal industry pays in the United States— then the government would have received revenues in excess of US$240 billion since 1872. This represents a massive subsidy to the mining industry, albeit an indirect one. "These are our lands they're exploiting," said Rodriguez, trembling with indignation. "They don't pay a cent in royalties, and they leave us the Summitvilles of this world. I think that's disgusting, and our congressmen and senators sit up there in Washington and do nothing about that law. I think that's the biggest rip-off the American public is facing today."

Actually, it is one rip-off among a whole constellation of rip-offs, which involve public losses for private gain, where out of foolishness, or ineptitude, or a desire to satisfy certain privileged constituencies, governments short-change the taxpayer and, at the same time, encourage businesses and individuals to destroy or damage the environment. In the awkward jargon of the day, we are talking about perverse subsidies: subsidies that are bad both for the economy and for the environment. That is what this book is about, and it is worth recounting briefly its genesis and purpose.

A few years ago, I was contacted by Lloyd Timberlake, a laconic Georgian whom I first worked with during the 1980s when he was writer-in-residence at the London-based International Institute for Environment and Development (IIED). He wanted to know whether I was willing to write a popular book—not, he stressed, an academic treatise or a jargon-laden diatribe—about subsidies and their malign impact on the environment. He explained that he was now working for AVINA, a charitable foundation based in Miami and established by Stephan Schmidheiny. One of the world's wealthiest businessmen and a leading figure in the World Business Council for Sustainable Development, Schmidheiny fervently believes that if there is to be a pleasant and sustainable future, then business has a key role to play. AVINA was concerned about the influence of government subsidies, and it had commissioned a report by two economists who worked for the Institute for Research on Public Expenditure in the Netherlands. The title of the report— *Addicted to Subsidies: How Governments Use Your Money To Destroy the Earth and Pamper the Rich*—gives a good indication of its content and flavor. Timberlake implied that many of the relevant facts and figures were here, but this was a specialist work, and unlikely to be read outside the narrow ghetto occupied by environmental activists and natural resource economists. The people who really needed to be alerted to the ways in which governments subsidize environmentally destructive activities were the people who were unwittingly paying for them: the taxpayer.

What was needed, suggested Timberlake, was a book that told real life stories about the men and women at the sharp end of the subsidy business, about people such as Ignacio Rodriguez and the small farmers of the San Luis Valley who were getting a raw deal, and about their adversaries, the corporations and individuals who were cashing in on government largesse. Timberlake thought I might be the right man for the job. Over the past two decades I had written several travelogues, mostly about journeys I had made through Africa, the Indian sub-continent, and parts of Europe, and I had also spent a good chunk of my life investigating environmental issues. I was sympathetic to environmental concerns, but sufficiently skeptical about some of the claims made by environmentalists not to be considered "one of them." I had also seen enough of industry and government not to be easily duped by their sophisticated public relations machines. In short, Timberlake believed that I would approach this subject with an open mind, and with the intention of entertaining the reader as well as myself.

I set about the task by reading around the subject, and I did so with a growing sense of disbelief. Governments around the world spend a significant chunk of public finance, measured in hundreds of billions of dollars a year, on subsidies. Four sectors alone—farming, road transport, and the energy and water industries—receive subsidies in excess of US$700 billion, which is roughly what the world spends on the arms race each year. By the time I was half way through the report that Lloyd Timberlake sent me, I could scarcely comprehend the scale of spending. It was like staring up into the night sky and reflecting on the fact that we can see the light of tens of thousands of stars that ceased to exist thousands of years ago. The sheer scale of the subsidies seemed almost as mind-bogglingly incomprehensible as the idea of infinity, or of light years as a measure of time. Once I had read that report, I turned to others, all of which told an extraordinary story of profligacy and destruction.

Once I began to get my eye in on the subsidies issue, it became increasingly clear that government programs were riddled with economic absurdities. Soon after I had accepted Timberlake's offer—AVINA kindly agreed to fund this book—I came across an illuminating piece in a UK newspaper. It told the story of a Welsh farmer who for several years had received the equivalent of, a US$24,000 subsidy from the European Union (EU) to grow 100 tons of flax, which he was then ordered, as a condition of the subsidy, to harvest and destroy, there being no readily accessible market for his crop. One year he decided to save himself the cost of harvesting, and do his soil a favor, by plowing the crop in. European Union inspectors—also funded by our taxes—learnt of his actions and took him to court, where he was prosecuted for obtaining a subsidy by deception. The judge described the case as "scarcely believable."

This led me to reflect on my own experiences of the farming world. During the early 1970s I worked on a mixed arable and dairy farm in the north of England, and as a farm student I was frequently entrusted with the least skilled and most dreary tasks: forking bales of hay, unloading wagons of fertilizer, shoveling cow muck, and the like. One of the less cerebral activities involved ripping out hedgerows, and during the course of a month I and a couple of others destroyed and burnt great lengths of hedgerow, which for two centuries or more had been home to a wide variety of plants, birds, and small animals. In doing so we turned half a dozen small, irregular fields into one large 100-acre field, thus gaining extra land on which to grow crops. From now on the business of plowing, sowing, and harvesting would also be much easier, and the farm more "efficient."

Much of this—like many other practices whose purpose was to make farming more efficient—was paid for by a government subsidy. Hedge removal subsidies no longer exist: over a short period of time they led to the rape of a beautiful landscape, to the loss of wildlife habitat, and—eventually—to a sense of outrage among the public. Nowadays farmers in some areas are paid a subsidy to plant hedgerows instead, which means that today's taxpayers are obliged to make amends for the damage done by activities that were paid for by yesterday's taxpayers.

In the days before I began my journey around the world of subsidies I heard of many other programs that, like the hedge removal and flax subsidies, seemed scarcely believable. Take, for example, the German coal industry. It would be cheaper for the government to close down all of its coal mines and give the miners full pay for the rest of their lives than to do what it does: throw billions of dollars a year at an inefficient and polluting industry. Many agriculture policies make just as little sense. In Europe, farmers are paid a subsidy to produce too much sugar; taxpayers must then pay to store the surplus sugar, or provide subsidies to export the sugar, which then depresses world market prices and puts peasant sugar producers in the Caribbean out of business.

Agricultural policies in the United States are at times equally absurd—a fact that was vividly brought home to me during my travels round the country. During 1999, farmers in the American Midwest were receiving subsidies to compensate for poor prices—prices that stemmed from farmers producing too much produce in the first place. At the same time, farmers in the East were being given subsidies to compensate for producing too little. So whether they lost crops to forces of nature, or saturated the market, they still got a slice of your money. What is more, the subsidies were often leading to higher, not lower, prices for the consumers, and most went to agribusiness, not to the small family farms for whom many subsidies were originally designed.

This Alice in Wonderland world of topsy-turvy economic practice was memorably foretold in Joseph Heller's *Catch-22*. Heller wrote of a farmer who received a subsidy not to grow alfalfa, so he bought more land on which not to grow alfalfa, and with the money he received for not growing more alfalfa, he could buy more land still, on which not to grow… and so on. Forty-odd years ago, this seemed like good comic writing; now, it is uncomfortably close to the truth. Heller's subsidy, like many that feature in this book, was conceived with what seemed like good intentions: in this case, to reduce the amount of alfalfa that was flooding the market. Similarly, the 1872 Mining Law was a perfectly sensible piece of legislation in the days when young men were being encouraged to go West. Now, it is absurd.

Lloyd Timberlake and I agreed that my travels should concentrate on the United States. There were two obvious reasons for this. For one thing, the United States is the world's largest and most significant economy, and it probably accounts for around one fifth of all perverse subsidies. For another, there has been a well-orchestrated campaign in the United States to curb the spending of public money on projects that harm the environment. Every year since 1993, Friends of the Earth (FoE), Taxpayers for Common Sense, and the United States Public Interest Research Group (PIRG) have published a

Green Scissors report. The 2001 edition highlighted 74 allegedly environmentally damaging projects or programs that were set to cost the taxpayer an estimated US$55 billion. If the government were to cut these, claimed the Green Scissors campaigners, that is what it would save. The campaigners estimate that they have helped to save more than US$24 billion since 1995, this being the value of programs cut by the federal government, partly or entirely as the result of the Green Scissors campaign. This is certainly better than nothing; but US$24 billion over four years is the merest drop in the subsidies ocean.

On reflection, I now realize that there was another good reason why it made sense to concentrate on the United States. The system by which corporations and individuals buy political favors is transparently open. One hesitates to use the word corruption, but the contributions made to political campaigns by big business clearly do corrupt the political process. Subsidies can be bought, and the retention of subsidies ensured, by the simple process of channeling dollars into the coffers of politicians. During a five-year period in the 1990s, the oil industry contributed over US$25 million to congressional candidates, the coal industry over US$20 million, and the timber industry US$16.5 million. It was, it seems, money well spent. During the same period subsidies to the oil industry amounted to around US$300 million, to the coal industry a staggering US$1.5 billion, and even more to the timber industry. The winner here is corporate America; the losers are you, the taxpayer. Legislation to reform the Campaign finance system was passed by the House of Representatives and the Senate in the Spring of 2002. This will take some of the money out of politics—but much remains.

The Subsidy Scandal provides an outsider's view of America. Judging from the reactions of many people I met, this may be no bad thing. "Oh, that's great," exclaimed a young woman from FoE when I explained what I was doing. "You're going to do a de Tocqueville on the environmental movement!" Well, I am not de Tocqueville, nor is this book in any way comparable to his magnificent *Democracy in America*, but I hope that I have approached my task with an inquisitiveness, and at times a sense of wonder, which owes something to my being a stranger. Although this book is about subsidies and the damage they do, it also explores the underlying values that influence the way Americans use their land and their resources. To appreciate why vast quantities of money are wasted on subsidies, one needs to understand the motivation and attitude of those involved—whether they are the politicians who sign the checks, the businesses and individuals who gratefully cash them, or the broad and at times fractious coalition of environmentalists and fiscal conservatives who so hotly oppose them.

My journey began at the Washington, DC, offices of FoE, one fine spring day when the sun shone from an ice-blue sky and the banks of the Potomac were dusted pink with cherry blossom. During the course of a week I visited the offices of many other organizations. Some were concerned with specific issues, such as mining, or saving the Alaskan rainforest. Some had a broader environmental agenda, and others were more concerned with economic prudence than saving landscapes and wildlife. But all, whether they belonged to the left or right of the political spectrum, or professed no allegiance at all, shared a common goal: they wanted the government to slash its perverse subsidies. This would be good both for the taxpayer—agricultural subsidies alone cost an American family of four around US$1500 a year—and for the environment.

It soon became obvious, as I tramped around the streets of the capital, that I had to establish precisely what sort of subsidies I should look at. The director of natural resources studies at a libertarian think-tank told me that public education was a subsidy. Some would dispute that. However, this book is not concerned with welfare subsidies, such as those that go towards education, health, support for the arts and the direct alleviation of poverty; rather, it is about economic subsidies, and especially those that are channeled towards businesses and individuals who directly or indirectly exploit natural resources: timber and mining companies, farmers, fishermen, energy extractors, road-builders, and so on. In the broadest terms, a subsidy is any measure instituted by governments that keeps prices for producers above the level they would be in a free, undistorted market, or prices for consumers below it, or which helps to reduce costs for either producers or consumers.

There is, as one US government report concedes, a dizzying array of subsidies, ranging from price supports to tax exemptions and low-cost loans, from exemptions from environmental laws to the provision of cheap services and import tariffs. One way or another I have sampled all of these, and while all but one of the stories in this book come from the United States, I have tried to set them within a broader global context. Politicians the world over seem to have an extraordinary fondness for spending their citizens' hard-earned taxes, and many of the stories here could equally well be told, with minor variations in place and personnel, for other nations with broadly similar economies to the United States.

The Subsidy Scandal is supposed to entertain as well as enlighten, and I have chosen the stories with this in mind. The space given to each—whether it relates to logging subsidies in Alaska, grazing subsidies in New Mexico or road-building in West Virginia—does not necessarily reflect the relative scale of government spending. Worldwide, subsidies to the automobile probably exceed forest subsidies by a factor of ten, but I have devoted as much space to Alaskan forests as to West Virginian highways. Likewise, energy subsidies dwarf fishing subsidies, but I have given the fishing story as much space as the chapter on energy subsidies. What matters, above all, are the principles involved. Why do governments persist in wasting our money? Why do politicians curry favor with environmental destroyers? Why are so many businesses addicted to subsidies? The government program to kill coyotes and other predators of livestock in the American West may only cost US$10 million a year, but it tells us as much about the way governments work, and about the ways in which politicians can be influenced, as the tax breaks given to the oil and gas industry—which in the United States are said to amount to around US$500 million a year.

When I embarked on my journey around the States, it never occurred to me that subsidies would be making daily headlines by the time this book went to press. In the months leading up to the November 2002 mid-term elections, the Bush administration performed two acts which made every free-trader and subsidy-slasher shudder with horror. The first was to slap tariffs of up to 30 percent on steel imports. The administration hoped that protecting US steel manufacturers from foreign competition would be a vote winner for the Republicans in Ohio, Pennsylvania and West Virginia. President Bush then signed a Farm Bill that will raise agricultural subsidies by 80 percent over the next ten years. This is a disaster for both the US taxpayer and for farmers elsewhere, whose produce will now have to compete with the cheap exports which the bill will generate.

Like the steel tariffs, the Farm Bill might even fail to achieve one of its main goals, which is to win votes in the mid-term elections. This is pork-barrel politics at its worst, and it has done immense damage to the Bush administration's reputation abroad. The United States, the great champion of free trade, is talking in forked tongues. Its message for the world is: you guys must liberalize your economies, get rid of your subsidies, tear down your tariffs. In the meantime, the United States is pursuing a protectionist agenda and bloating the subsidy bill. Mind you, US commentators are right to scoff at the outrage which wafts across the Atlantic: European nations are every bit as guilty of perpetuating the subsidy scandal as the United States.

My thanks go to all of those who helped and entertained me. Unlike the British, who are masters of obfuscation, American people are for the most part disingenuously open, and frequently outspoken, sometimes libelously so. The very nature of my travels meant that wherever I went, I spent time with people who were at loggerheads with one another: with ranchers and anti-grazing activists in New

Mexico; with dam-builders and dam-haters in California; with swamp-loving conservationists and swamp-draining farmers in Florida. Consequently, I often found myself liking people who detested one another. I have done my best not to be partisan. The environmentalists will not always be happy with my conclusions; neither will those who unconditionally oppose all subsidies. However, the people who come off worse in these stories are the politicians who have kept the subsidy gravy train rolling, and the individuals, bureaucracies and businesses who have made a living, partly or wholly, at the expense of the taxpayer. If they are offended by what I have written, I hope they will accept that my criticisms do not stem from personal animosity, but from the nature of what they do.

1

PULPING THE FORESTS

In 1867 William Seward, the US secretary of state, signed a deal in the remote settlement of Sitka that many Americans referred to at the time as "Seward's folly." He paid the Russians the sum of US$7.2 million for all their possessions in Alaska, and Sitka, hitherto the capital of Russian America, became the outpost of the US federal government in its newly acquired territory. As it turned out, Seward's deal amounted to something close to a gambler's dream: for a minimal pay-out, it was to yield vast profits. The Russians had plundered one natural resource—decimating the sea otter, bearer of an exquisite and highly prized pelt—but they had made only cursory use of the region's forests and minerals. Soon after the Russians left there was a gold rush, and a century later wealth-bestowing oil began to flow freely along the trans-Alaska pipeline. Indeed, Seward's perspicacity, or at least that of his government, led to the enrichment of both the US Treasury and the Alaskan people: each individual, from the most industrious to the least deserving, now receives an annual oil dividend of around US$2000.

These gains are in stark contrast to the financial losses that have resulted from federal forestry policy in the state. Over the past 50 years, billions of dollars of taxpayers' money have been used to help two major pulp companies exploit—and destroy—part of the world's largest temperate rainforest. The pulp companies have left now, but the timber program in Alaska continues to soak up at least US$26 million a year in federal subsidies.

Approach Sitka from the sea—we arrived in a converted World War II mine-sweeper—and it looks much as you might expect a small coastal Alaskan settlement to look. Salmon trollers and an assortment of other craft are moored in the harbor, and the waterfront is punctuated by ramshackle warehouses, government buildings, and the odd bar. It comes as something of a surprise to realize that the birds, that swooping by the score around the harbor buoys, and festooning the spruce trees like oversized Christmas decorations, are not gulls, but bald-headed eagles, and there are more surprises once you step ashore. Dominating the main street is St. Michael's cathedral, a good example of rustic Russian architecture, dripping with Orthodox icons and surrounded by shops selling Russian dolls and other baubles.

For such a small place—Sitka's population is less than 9000—there is a fine mixture of restaurants, bars, bookstores, and art galleries. Tourism has much to do with this; Sitka is on the itinerary of the major cruise liners, but the cosmopolitan nature of the town also reflects its social diversity. Around one third of the population are native Tlingit, some of whose ancestors drove out the Russians when they arrived at the end of the 18th century, and some of whose ancestors were killed a couple of years later when the Russians successfully reinvaded. Of the white population, the vast majority have come from elsewhere, with men far outnumbering women. Many of the former have the look and demeanor of frontiersmen. A good number see hard drinking as a discipline rather than a recreation, and provide the state's women with the opportunity to inform strangers that when it comes to getting a man, "the odds are good, but the goods are odd."

On the evening of our arrival, we were invited by members of the Sitka Conservation Society to a barbecue. Our party consisted of two staffers from the House of Representatives, two forestry experts, a gravel-voiced New York journalist, several people who worked for local conservation groups, and Richard Nelson, the celebrated nature writer and an inhabitant of Sitka. Together we had spent the past three days, at the invitation of the Alaska Rainforest Campaign, viewing some of the best and worst of the Tongass forest: magnificent old-growth and great swathes of clear-cut.

The conservationists were in good heart. Six years ago the waters of Sitka Sound were whisky-brown, polluted by effluent and the decaying logs that constituted the raw material for the Alaska Pulp Corporation's mill a couple of miles out of town. The mill had closed down in 1993 and over the years the water had cleared. "When it shut down," recalled Richard Nelson, a slim, gray-haired, bespectacled figure with a soft voice and a gently humorous manner, "it was like a miracle for us." Just as significant, in many ways, was the recent announcement by the US agriculture under-secretary of a major reform in Tongass forestry policy. The under-secretary had reduced the amount of timber that the Forest Service could offer for sale by around one third, and declared 18 areas, which formerly could have been subject to logging, as inviolate, thus adding another half a million acres to the stock of wilderness.

The conservationists were understandably delighted by this change in policy: the new plan gave them more than they had dared hope for. Not surprisingly, it was condemned by the Alaska Forest Association, which represents the logging industry. Its executive director, Jack Phelps, told the local press that it was "a slap in the face to the communities, the workers, and the citizens of Alaska."

Stan Filler, the mayor of Sitka, agreed. I rang him shortly before we left for the barbecue and he suggested I meet him the following morning at Ernie's Old Time Saloon. "I own it," he explained peremptorily. A short man with a powerful physique and a weathered, lived-in face, the mayor opened the door at precisely 8 o'clock. By the time he had returned behind the bar four people had climbed onto stools to await their early morning fixes below the stuffed heads of black bear and other hunting trophies. The mayor dispensed a Bloody Mary to a woman whose shaking hands told their own story, and a couple of beers to two elderly men. Then he poured us some black coffee and we headed for a booth.

The mayor thought that the community was split 50:50 on the logging issue, and his sympathies lay with those who wished to see large-scale logging continue. "The Clinton administration screwed the whole thing up," he explained as he lit a Marlboro Light. "The pulp mill up the road had a 50-year contract, but then the government made it impossible for them to carry on. The whole idea of these contracts was to supply year-round jobs. That's what the region needed after World War II, and that's what it still needs today." He said the demise of the mill had been a disaster for the town. Of those who lost their jobs at the mill, around 150 left Sitka immediately, and there had been a knock-on effect on many trades and local businesses. "There was a lot of denial, too," reflected the mayor. "A lot of folk said, 'Well, hell, it'll open up again soon,' and they spent their severance money. It's caused a lot of pain. I see guys at Alcoholics Anonymous and I say, 'What the hell are you here for?' They used to have just a six-pack on the weekend, but things changed after they lost their jobs, and they tell me: 'I started drinking and my health's suffering and my liver's shot to pieces and the doctor says that if I don't stop. . .'." The mayor lit another cigarette. "There've been suicides and broken marriages too on account of the mill closing."

At a recent public meeting the mayor told the audience that there were three words he didn't want to hear: "they", "them", and "those". "Because this is *our* town, and one way or another we've all got to live together," he explained. But the community remained divided, and there was a great gulf in outlook and lifestyle between the sort of people I had spent the previous evening with—mostly professionals, and many here to live out a comfortable retirement— and the clientele who frequented the mayor's bar. To paraphrase Charles Dickens, it was the best of times, it was the worst of times, it was the spring of hope, it was the winter of despair. Which view you took depended upon who you were and how you made a living—or didn't. "It's not about timber," said the mayor at one point. "Hell, no, timber's just something we've got something of. It's about jobs."

It is also about the use of public money. The cynic may say that many subsidies are designed by politicians with the sole purpose of enriching themselves and their buddies, or as a form of patronage to help them retain power. However, some subsidies are designed with perfectly honorable intentions, the most obvious one being to foster economic development and create employment opportunities. This was the explicit aim of the subsidized forestry program in the United States, particularly in remote regions such as Alaska. But had it worked? That was what I had come to find out.

*   *   *

Our journey towards Sitka had begun in the state capital of Juneau. Here, our disparate band gathered one chilly June afternoon, and we took an ancient seaplane south to Tenakee Springs, where we swapped places on the 12-berth *MV Observer* with four members of the House of Representatives and their spouses, among them the Californian Democrat George Miller, now a celebrated figure in the battle to curb perverse subsidies. The Alaska Rainforest Campaign had brought the congressmen up to the Tongass to enlighten them about forestry issues, in the hope that over the coming years they would argue the conservation cause in Washington, DC, where they were now returning. Unfortunately for the politicians, it had rained almost incessantly during their visit. This particular evening the clouds had cleared, giving us long views over the snow-capped peaks that reared above the green flanks of Chichagof Island. Before dinner we went ashore and headed along Tenakee Spring's main street, a rutted and mossy dirt lane, past log houses perched above the water on barnacle-encrusted stilts, past the small harbor, and onto a wide expanse of rocky shore. Rooting around among the seaweed on the shore's edge were a grizzly bear and two half-grown cubs. We approached to within a hundred yards or so, then the hairs stood up on the back of my neck, prudence got the better of us, and we swiftly made our way back to the first of many grand meals on the mine-sweeper.

The next morning was fine, too, and after breakfast we headed out into Chatham Strait, a broad channel of water separating Chichagof and Admiralty Islands, then sat in the galley while Matt Zencey of the Alaska Rainforest Campaign and Richard Nelson gave us a potted history of the Tongass. Every now and then, they would break off and motion at the passing landscape: to a mosaic of clear-cuts, to stands of pristine forest, and on one occasion to a small flotilla of humpbacked whales.

The Tongass covers 16.9 million acres of south-east Alaska. Around two-thirds of this is rock, ice, and scrub forest of no commercial value. The remaining third has some commercial potential, but only about 4 percent of the Tongass is—or was—what Zencey described as "really good stuff." Unfortunately, it is good stuff from everybody's point of view, whether they are loggers, conservationists, fishermen, or recreationists. The loggers covet it as it contains the highest volume of the best trees, and has the potential to yield the greatest profits. The conservationists love it as it is the habitat that best suits bear, deer, and other wildlife. Commercial salmon fishermen are almost as fervent about the need to protect such forests, as the survival of the salmon depends upon the integrity of the riverside habitat. And hikers, hunters, and sports fishermen see the old-growth forest as one of the state's key attractions.

For hundreds of years the Tongass provided timber in modest quantities to the native people, and during the first half of the 20th century it was the source of timber for various small enterprises. However, all of this was to change with the arrival of industrial-scale logging after World War II. "There was a deliberate policy decision," explained Zencey, who with his black beard and luxuriant hair looked like a grizzled explorer. "The federal government decided to turn old-growth trees into jobs." In the 1950s the government signed 50-year contracts with two logging corporations. An American corporation set up shop in Ketchican; a Japanese corporation in Sitka. The idea was simple. The Forest Service would provide the companies with an adequate quantity of timber. High-quality cedar would be sold as whole logs; spruce and hemlock would be turned to pulp and to pulp products. Within a short space of time the price-fixing between the companies and the Forest Service squeezed the smaller concerns out of business, and Ketchican Pulp Company, a subsidiary of the Louisiana Pacific Corporation, and the Japanese-owned Alaska Pulp Corporation, had a virtual duopoly.

Neither company had to bid for timber competitively, as they would have done in a free market. The price was determined by the Forest Service, and according to the conservationists it was fixed at a level that ensured the profitability of the logging operations. "It was as though the Forest Service had an out-of-body experience," ventured Amelia Jenkins, a member of our party and a director of Forest Service Employees for Environmental Ethics. "They'd jump out of the service and become a logger. And they'd look at all the logging costs and the profit margins and work out a price that would suit them if they were loggers." Over the next 40 years subsidy-driven forest policy led to the felling of around 400,000 acres of prime Tongass forest.

Although most of the forests we saw were owned by the state and managed by the Forest Service, some were managed—and at times mismanaged—by others. Soon after we set off we passed an enormous swathe of cleared land, many miles long, stretching from high up the snow-streaked mountains right down to the waterside. It struck me as strange that neither Zencey nor Nelson commented on this spectacular piece of devastation. Why, I inquired, had nothing been said about it?

Zencey explained, somewhat sheepishly, that this was one of the many clear-cuts that had been carried out by the native corporations on land they acquired under the 1971 Native Claims Settlement Act. Nelson conceded that the native corporations had devastated vast areas of forest, but said that the white community was reluctant to interfere in their affairs. "For me it would be unethical if we did interfere," he said. Later, when I asked Buck Lindekugel, the conservation director of the South-East Alaska Conservation Council (SEACC), why his organization said so little about the native corporations' plunder, he pointed out that they were logging on their own private land. The Forest Service, in contrast, managed public lands using public funds: this was an issue in which all citizens could rightly take an interest, and where it was possible to effect change. Point taken; but had any non-native communities ravaged their forests with the same blunt enthusiasm as the native corporations had theirs, you can be sure there would have been a national outcry.

If Lindekugel had been born an animal he would have been a walrus. A large man with shaggy blond hair, a droopy mustache, and a lugubrious manner, he delivered his opinions about the Forest Service and the loggers in an insistent monotone. In his view, the mythology of superabundance—the belief that nature would always provide—had plagued the United States throughout its recent history. In Alaska, the mythology still held sway. "The fact is," he said, paraphrasing Barry Lopez, a literary friend of Richard Nelson, "there are not enough resources on public land to make everyone rich, and trying to do so will impoverish all of us." It was this belief that there are limits to growth that led to the foundation of SEACC. Appalled by the scale of forest loss in Alaska, SEACC began to agitate against government policy during the 1970s, and it was partly as a result of its campaigns that the federal government passed the Alaska Lands Act of 1980. Under the act, most of Admiralty Island was declared a national monument and large portions of the Tongass were designated as wilderness, and therefore off limits to the loggers.

In the view of Ted Stevens, a local politician and now chairman of the US Senate Appropriations Committee, the new law went far too far in favor of conservation and he squeezed a remarkable concession out of the government: this was the Tongass Timber Supply Fund, which mandated an annual subsidy of US$40 million to the Forest Service in Alaska. Over the next decade the fund poured US$400 million of public money into Alaska, and the sole beneficiaries were the logging companies and their handmaiden, the US Forest Service.

SEACC's strength, then as now, owed much to its breadth of interests. A coalition of local organizations representing commercial fishermen, hunters, native Alaskans, small-scale timber operators, tourist enterprises, and conservationists, SEACC was able to mount a vigorous and credible campaign during the 1980s against the fiscal absurdities of the supply fund. Through rigorous research it was able to establish that the federal government was not only wasting public money, it was channeling it into activities that were having serious implications for the native wildlife and, consequently, for the local people who depended upon the wildlife. In places such as Sitka, three out of four families have at least one member who fishes for subsistence, and almost as many families are involved in hunting for food. Indeed, if south-east Alaskans, who number around 70,000, were to replace wild foods with store-bought equivalents, they would be paying an estimated US$22–US$35 million more for their food each year than they do.

On our journey we saw a great deal of wildlife, almost always in areas where the forest was in good health. Now and again we would anchor in a sheltered bay, prize ourselves away from the dining room and head ashore in an inflatable dinghy. One day, as we fished a rushing torrent for trout, we saw a herd of black-tailed deer. On another, we wandered deep into the forest. Below the majestic trees the ground was luxuriant and damp, with mosses and liverworts and delicate grasses forming a thick spongy carpet. Richard Nelson showed us signs of deer, and we inspected, apprehensively, the fresh droppings of a grizzly bear.

Without looking too hard we saw a dozen or so grizzlies during our trip. "As far as we can tell," explained Nelson over a halibut dinner, caught earlier in the day by one of our party, "there are around 30,000 grizzlies in Alaska." This is the only state in the union where they are still plentiful. In the lower 48, there used to be 100,000 or so grizzlies; now there are fewer than 1000, and a further onslaught on the Tongass would almost certainly lead to their decline here.

On my return to Juneau, I mentioned to Lana Shea Flanders, an ecologist with the state's Department of Fish and Game, that I had heard conflicting stories about deer and old-growth forests. The mayor of Sitka, Stan Filler, had said that clear-cutting provided deer with more space and fodder and that their numbers had increased over recent years. Richard Nelson disagreed. He cited studies conducted on Vancouver Island that found a 50–75 percent decline in deer harvest in heavily logged areas. He said that in winter, when there was heavy snow, the trees helped to shelter the ground fodder upon which the deer depended; without trees they would starve. Lana said that Nelson was right: there was no doubt that the widespread clear-cutting of old-growth would eventually affect the deer. "If we have more mild winters," she said, "you won't see the expected decline in deer following logging until the clear-cut is around 25 to 30 years' old, and new growth shades out the deer food." However, she predicted that a harsh winter with heavy snowfall would lead to a significant deer die-off in logged areas.

Her department was also worried about the effect of forest roads and logging camps on grizzly bear. "What we've noticed is that where there are roads you get uncontrolled hunting and there tends to be an increase of 'defense of life and property' kills." Alaska is also important for the wolf. Sadly, said Lana, a decline in the moose population had led to an increase in wolf killing, as humans saw the wolf as a direct competitor for moose. Lana feared that the decline in the deer population could have a similar impact upon wolves in south-east Alaska. "Wolves depend on deer," she explained. "Deer depend on old-growth forests. Humans depend on deer. If the deer start declining because of logging, then humans will go after the wolves." There would be little logging were it not for the subsidies, so the subsidies could well turn out to be the wolves' nemesis.

To return to the history of Tongass subsidies: the 1980s was a time of financial madness, heavy logging and growing conflict between the Forest Service and the corporations, on the one hand, and the conservationists and their allies on the other. Pressure from the latter and outrage in Washington, DC, at the annual US$40 million subsidy forced Congress to re-evaluate the role of the Forest Service in the Tongass. To cut a long story short, this led to the Tongass Timber Reform Act of 1990. The act abolished the annual mandated US$40 million subsidy, modified the 50-year contracts and designated a further million acres of old-growth of forest as wilderness, which could not be logged.

Astonishingly, logging levels actually increased over the next three years: the Forest Service was in no mood to change its ways. But change has come. In 1993 the Alaska Pulp Corporation closed down its operations in Sitka; in 1997 Louisiana Pacific closed its pulp mill at Ketchican. "They were basically done in by economics," reflected Brian McNitt of the Sitka Conservation Society, when I visited his office across the street from Ernie's Old Time Saloon. "They'd cut the most valuable timber, and the market price for pulp was falling." The subsidy gravy train had also slowed down; in fact, had it not been for the subsidies these corporations would probably never have come to Alaska in the first place. The final *coup de grace* was delivered by the environmental agencies, which were demanding that the mills reduce their emissions and waste. "They were here for as long as they could make a profit," explained McNitt. "And they made huge profits. I've heard from insiders that they had no real plans to be here for a long time." Neither had bothered to update their machinery, and both mills operated their workforce without long-term contracts. The Japanese corporation agreed to give its mill to the town of Sitka, with the proviso that it would not be liable for any unforeseen pollution event that might arise from the mill's clearance and reclamation. Louisiana Pacific did even better: in order to terminate its 50-year contract for timber, scheduled to end in 2004, the government paid it a large compensatory sum.

During the 1990s, an assortment of other subsidies related to forestry issues also wended their way to Alaska, besides those already in existence. Most notably, communities affected by the closure of the pulp mills were awarded US$110 million, a sum of money extracted from the government by Senator Ted Stevens. Stevens saw this as disaster relief; organizations such as SEACC saw it as a sum of money that would help communities in places such as Sitka and Ketchican make the transition away from industrial logging to a more sustainable pattern of development. "Others," suggested Buck Lindekugel in a droll aside, "simply saw it as Ted Stevens holding the country to ransom." The Senate Appropriations Committee, of which Stevens was chairman, put forward a bill for the Tongass appropriation in 1996 that included an amendment which would have increased the amount of timber that the Forest Service could sell. President Clinton refused to sign, and as a result the government was temporarily closed down. Eventually, Ted Stevens agreed to withdraw the amendment in return for US$110 million of taxpayers' money.

A year later, in 1997, Congress allocated the sum of US$130 million to Louisiana Pacific to settle the contract dispute. The following year Congress spent a further US$12.5 million on the Tongass. This was allocated to the Forest Service in order to prepare timber sales for the following year. As it happens, less than 20 percent of the timber offered by the Forest Service was sold. Why? Because the downturn in the market meant that it made little financial sense to buy standing timber in Alaska—at least, during this particular time.

*   *   *

It was not long before I realized that behind every subsidy in the United States there was a politician or a band of politicians who made it their business to ensure that the subsidy survived. In the case of Alaska, it was the exceptionally well-placed delegation, consisting of Ted Stevens, Frank Murkowski, and Don Young. Murkowski was chairman of the Senate Energy and Natural Resources Committee, and Young was chairman of the House Resources Committee. The delegation was greatly admired by many Alaskans, and when I mentioned their names to Stan Filler he said that they should be praised for their efforts. "The Alaska delegation are personal friends of mine," the mayor told me in his bar, "and they've done a lot of good things for our state." The following day he was heading off to Washington, DC, to talk to the delegation about Sitka's needs. He said he was going to take a cookbook for Frank Murkowski's wife.

The environmentalists, on the other hand, had nothing good to say about the delegation. While they applauded those whose views coincided with theirs, they vilified those whom they perceived as acting against their interests. They talked in reverential terms about Congressman George Miller, who opposed what they saw as perverse subsidies, and attached epithets such as "evil" to men such as Ted Stevens. Their attitude was: "If you aren't with us, you're against us." But if the environmentalists were at odds with the sort of people who frequented places such as Ernie's Old Time Saloon, they were also out of sorts with their sometime allies in the war against perverse subsidies: the libertarians and fiscal conservatives.

The organization that represents fiscal conservatism in its purest form—the 40 percent proof version—is the Cato Institute, a think-tank based in Washington, DC. I was pointed in the direction of Jerry Taylor, Cato's natural resources director, by the delightful Jill Lancelot, a founder member of Taxpayers for Common Sense. Lancelot had been campaigning for years against government waste and helped to found the Green Scissors campaign against perverse subsidies. She said that Taylor was a must-see: he was clever, charismatic and opinionated, and he would provide me with a very different slant on the subsidy issue, and the role of politicians, from the one that I would hear from environmentalists.

A handsome, sharp-suited, snappy-talking man with the manner of an aggressive attorney, Taylor began our meeting in his DC office by explaining that he and his colleagues believed in classical liberalism: they were the intellectual heirs of Locke, Cobden, and Bright, three great apostles of free trade and limited government. "Government does a lot of dumb things in all aspects of life," he said expansively, "whether they're regulating drugs, trying to grow crops, or trying to run a justice system, and there shouldn't be any surprise that when they try to foster economic development, there are often bad environmental consequences." Were it not for government subsidy, he said, there probably wouldn't have been any big dams in the American West. The sugar program was ridiculous; so were coal subsidies. And as for roads such as Corridor H, which I told him I intended to visit, the idea that remote West Virginia needed a four-lane highway was absurd. Markets, he said, not government bureaucrats, were the most efficient means of allocating resources. "I don't believe in subsidy of any kind," said Taylor. But getting rid of them was exceptionally tricky, not least because of the nature of the political process. "I have what some would say is a jaded view—I'd say it was a mature view—of the nature of government," he explained. "Politicians do not decide on policies due to a careful weighing of evidence, of reports. Policies are based on a political need of appeasing constituents who hold their careers at their whim." Politics, in other words, is all about currying favor with those who can help you retain power.

Peter VanDoren, an economist who had come to the Cato Institute via Yale and various academic institutions, expanded on this when he joined us. "How come majorities often lose?" he mused rhetorically. "They lose because politics is a series of trades across votes. You vote for me on this and I'll vote for you on that. The economy is worse off, but everyone gets re-elected." So politicians in the Midwest with a strong agricultural constituency will vote "yes" to the sugar program, because Florida's sugar-funded politicians will vote to continue the Midwest cereal subsidies. The Western

Republicans—much detested by the environmentalists—in Colorado and New Mexico will vote to continue timber subsidies in Alaska, because Alaska's politicians will vote to continue ranching subsidies in their states.

Had I heard of Senator William Proxmire of Wisconsin, asked Taylor? During the 1980s the senator had established the annual Golden Fleece Award to shed light on what Taylor described as "pork-barrel economic mischief." The worst cases of perverse subsidy won the award—one year it went to the Tongass timber program subsidies—and this proved to be an eye-catching way of publicizing the waste of public money on often destructive programs and projects. "But there was one exception to this rule," said Taylor with an expression which foretold a good punch line. "Proxmire thought that there was nothing wrong with the subsidies that went to his constituents, Wisconsin's dairy farmers. To him, milk price support was sacred." So even the fiscal conservatives on Capitol Hill would exempt themselves from the strictures they happily applied to others.

We then talked about Corridor H, and the influence of Senator Robert C. Byrd, who for years had been pressing for federal funds to build the highway across the Appalachians. "From the senator's point of view, redistributing taxpayer dollars to West Virginia is what he's supposed to do," said VanDoren. "And he does it very well." Let us say, he continued, that a "good government" candidate was to run against Senator Byrd, a candidate who espoused a Cato platform: "He's against subsidies. Everyone should pay the full costs of the stuff they use. That sounds good. It's very American. Senator Byrd will kill that person in elections. Why? Because West Virginia loses if you go against Byrd." This is precisely what happened during the last election to a Republican in South Carolina who suggested to the electors that America's founding fathers hadn't set up America so that its people could loot one another. "He was crushed," said Taylor, "because if he had won, then everyone else would have looted from the state of South Carolina."

Whether this is a jaded, or mature, view of American politics depends, I suppose, upon what you expect of politicians. What is clear is that the Alaska delegation has been supremely successful at looting the public purse. This was appreciated by the majority of Alaskans, who persistently voted for the same three individuals. You may not like the outcome, or the system, but there is nothing undemocratic about politicians looking after their own.

However, there is one aspect of the political process that is profoundly undemocratic, and if it existed in the Congo or Cuba, it would be held up as a symptom of their backwardness by governments of the Western world. The practice of attaching what are known as riders to spending bills has played a significant role in the retention and promotion of perverse subsidies, not least in Alaska.

Each year, Congress has to pass 13 appropriation bills to fund government business. If any one bill fails to pass, then government shuts down. It simply would not do for the largest and most prosperous nation in the world to go rudderless for any length of time, and consequently the administration uses its power of veto sparingly. This is a cue for some sharp practice by the law-makers, who frequently attach riders to bills in the knowledge that they will not be subject to debate or rigorous scrutiny. Astonishingly, these riders often have nothing whatsoever to do with the appropriations. For example, while I was in Washington, DC, during the spring of 1999, an appropriation bill to release funds for US military involvement in Iraq and for flood relief in Central America was accompanied by a rider whose purpose was to increase the timber allocation in Alaska.

In a report entitled *Riders Ransack Our Resources*, the US Public Interest Research Group (PIRG) analyzed the intent of the riders attached to the 1998 Interiors Appropriations Bill, which funds the various agencies, including the Forest Service, who are responsible for managing federal lands. The House of Representatives' version of the bill had what PIRG described as three anti-environmental riders; the US Senate version had at least 22. Ten of the Senate riders and three House riders, had they been enacted, would have cost taxpayers in the region of US$180 million and done considerable damage to the environment. Nine of the Senate riders were related to forests, and in every case they were designed to further their exploitation. One of these—it was later dropped—would have forced the Forest Service to allow the logging of more than double the amount cut in the Tongass during the previous year. Senator Stevens agreed to drop the rider in return for the US$12.5 million mentioned above and awarded to the Forest Service. I've yet to hear a sensible defense of the riders system, and I doubt whether there is one. But I detect a certain weariness among the editorial writers who regularly deplore the use of riders. It is almost as though riders are accepted as an unavoidable nuisance, like mosquitoes in the Alaskan summer.

*   *   *

Throughout the 1990s, the Forest Service was busy drawing up a management plan for the Tongass, and its eventual publication was rebuked by Senator Murkowski with the memorable remark that it was a "scientific insult, legal affront and economic crime." The *Washington Post* suggested that this was proof of its virtues. After much debate, a revised version was issued, and its contents came as a pleasant surprise to organizations such as SEACC and the Alaska Rainforest Campaign. Besides reducing the allowable harvest, and locking up a further half a million acres as wilderness, the plan increased the harvesting cycle from 100 to 200 years in 42 areas of importance to wildlife. The agriculture under-secretary's "Record of Decision" on the Tongass, or the 1999 ROD as it was known, was widely seen by the conservationists as a slap in the face to both the logging industry and the regional Forest Service.

So was that how Jim Caplan, the deputy regional forester, saw it?

"I'd say there's the full range of opinion in this office," he said warily when I saw him in Juneau.

"But are you happy with it?" I pressed.

"I have to say I'm prepared to implement it, because that's the nature of my job. It is a document that is implementable."

A quietly spoken man with a small mustache and spectacles, Caplan had about him the air of a well-adjusted academic. He was the sort of person it would be impossible to dislike and it was a pleasure to spend time with someone who was generous about those who were frequently critical of his organization. "I think the folks at SEACC are a wonderful bunch of people," he said at one point. "They are good advocates and they know what they want. That's a positive force." He added, however, that there were many more voices in the debate than SEACC's.

I told Caplan that Buck Lindekugel described the Forest Service as a "timber first agency." He maintained that although it had a remit to oversee a host of other activities, from wildlife conservation to the provision of recreational facilities, it was still primarily concerned with the sale of timber. Would the new ROD change his agency's approach, I asked. "We've already changed," he said bluntly. "In the early 1990s the agency restated its mission as ecosystem management. We've said we'll do everything we can to get rid of clear-cutting. The change is over. They're the ones, the guys at SEACC, who are having a hard time adapting." By this I presume he meant that SEACC was locked into confrontation mode, and was unwilling to adopt a more consensual approach to forest issues.

However, Lindekugel pointed out that the 1999 ROD would do nothing to halt the flow of subsidies to the Tongass. "We don't know exactly how much the subsidies amount to," he said, "because the Forest Service won't tell us what the figures were for 1998." He suggested that they would probably be of the same order of magnitude as they were in the recent past, around US$33 million a year. "The fact is they will carry on building roads left and right. That's what they do: they tame the wilderness. Regardless of the cost to the taxpayer, regardless of the fact that they can barely sell a stick of timber, they are still building roads." And this was a nonsense, said Lindekugel: why build new roads when there was over 10 billion board feet of timber—over 50 times the maximum allowable annual cut—adjacent to existing roads?

Caplan agreed that in the past much of the subsidy went on the building of roads. "At the moment we're not building much road," he added, "but in a few more years we might." At present the way the agency had planned its program assumed that the logging over the next few years would take place along existing roads. But the 1999 ROD anticipated more road-building in the future.

In fact, forest road-building has cost the taxpayer a fortune. According to the General Accounting Office, the US taxpayer contributed US$387.1 million to the building of forest roads during the five-year period of 1992–1997, and there are now over 440,000 miles of road on national forest land, sufficient to circle the globe 17 times. Until recently, the Purchaser Road Credit Scheme paid logging companies in trees for the roads they constructed. The supposition was that these forest roads, once built, were a public good, but this simply hasn't been the case for places such as the Tongass. The high cost of maintaining roads in a region with so much rainfall, and the lack of access to other roads, has meant that the forest roads are not the asset for recreationists, hunters, and others that the Forest Service has often claimed them to be. Congress stopped the scheme in 1998, but logging roads continue to be subsidized by the annual appropriations.

It is not just forest roads that soak up taxpayer dollars: an assortment of other forest-related subsidies must be factored into a balance sheet which shows that for every US dollar spent on the forest program less than 10 cents makes it way back to the Treasury. The US Forest Service Salvage Fund, whose purpose is to hasten the removal of dead, damaged, and diseased trees, finances around one third of the logging on national forests and is free from congressional scrutiny. The Forest Service has used the salvage sales to create a considerable fund—around US$170 million in late

1998—but has failed to pass any of this back to the US Treasury. According to the *Green Scissors* report, the fund provides local forest managers with an incentive to enhance sales by adding higher value green trees to low-value salvage material. The report also suggests that taxpayers are being relieved of a further US$50 million a year: this is the amount of money that the agency siphons off from the Knutson–Vandenburg Fund. The fund was set up during the 1930s to pay for reforestation of logged areas in the national forest system; but in 1997 49 percent of the fund was used on Forest Service overheads, not on replanting.

*   *   *

Shortly before I left Sitka, Richard Nelson gave me a copy of his book *The Island Within*, a poetic account of the time he spent on an unnamed, uninhabited island in the Pacific north-west. Nelson belongs to a long line of American transcendentalists—the best known, perhaps, is Henry David Thoreau—who believe that there is far more to nature than protoplasm and procreation, savagery and survival. *The Island Within* is a personal travelogue, and it reveals as much about Nelson and his beliefs as it does about his surroundings, even though these are observed in minute detail. What Nelson seems to hanker for is not a world from which man is entirely excluded, but one in which he makes more intelligent and benign use of nature; and he is much taken by the Inuit, with whom he spent time as young anthropologist, and by their relationship with nature. *The Island Within* is, above all, about one man's reverence for the natural world. After he shoots a deer, Nelson reflects on what he has done: "I whisper thanks to the animal, hoping I might be worthy of it, worthy of carrying on the life it has given, worthy of sharing in the larger life of which the deer and I are part."

I doubt whether this sort of thing goes down well in Ernie's Old Time Saloon, but it gives us a clue as to why Nelson, and many others, are prepared to devote so much time and energy to campaigning on environmental issues: diminishing nature, in their view, diminishes them. Of course, Nelson is fortunate in that he can make a living out of nature by writing about it. Likewise, many others I met were actually paid to work for the conservation cause. This is something that I believe they, and we, should remember when advocating measures that will lead to others losing their jobs.

Jobs were very much on the mayor's mind, and he was mourning their loss, when a small, dapper figure came rushing through the door of Ernie's Old Time Saloon and took a quick look at the early morning drinkers. "You want to see me?" asked the mayor. "No," replied the newcomer. He said he was looking for someone whose name was familiar to the mayor, and he waved a subpoena by way of explanation. "I threw him out of here years ago, and I've never let him back in," explained the mayor.

"You see," said the mayor as the bailiff's assistant departed, "that's one of the guys who used to work in the pulp mill. He earned good money at the mill—US$4500 a month—and now he and his wife are just scraping by." It was an all too familiar story, he said. The mill had provided jobs and brought prosperity to the area. There were other benefits that had flowed from their presence, too. "We've got a good transportation system," he continued. "Why? 'Cos of the mills up here. And they gave south-east Alaska stability."

There is no denying that the long-term contracts did, indeed, bring employment to the region—for a while. However, events since the closure of the mills simply do not support the mayor's contention that forest-related jobs are essential to the region's prosperity. During the early 1990s, as fears about the future of the pulp industry grew, there were dire predictions about the effect that mill closures would have on Sitka and Ketchikan. The McDowell group of economists forecast that Sitka would lose 923 jobs and that 1900 people would leave if the mill closed; this amounted to 21 percent of the workers and 22 percent of the population. The mill closed in 1993. By 1996, a mere 5 percent of the population had left and job losses were one quarter of those predicted. By

1998 the town had regained two-thirds of its lost population. A similar story can be told for Ketchican, which continues to prosper despite the loss of its pulp and saw mills.

Those who argue in favor of continuing the subsidies, and reviving large-scale forestry operations, on the grounds that they will provide work, ignore the fact that even during the early 1990s the forestry sector was not the major employer it was often claimed to be by the logging industry. By 1994 tourism and recreation in south-east Alaska provided 2771 jobs, compared to 2204 in the forestry sector and 1899 in salmon harvesting. Since then the number of forest-related jobs has steadily declined; the number of jobs in tourism and recreation has dramatically increased. Every year 700,000 people—approximately ten times the local population—visit south-east Alaska and a growing proportion come to watch or shoot wildlife, to fish, and to recreate in the Tongass. Figures provided by the Forest Service confirm that in economic and employment terms, logging is one of the least significant activities in national forests in the United States: recreation is far more important.

The fact that forest subsidies failed to stimulate long-term economic development in Alaska should come as no surprise: a similar story can be told for many other parts of the world. Take Chile, for example, where the government introduced plantation subsidies that had the opposite effect than intended. Over a short period of time they led to three large enterprises assuming owner-ship of 70 percent of all of the country's plantations, and they resulted in the expulsion of peasants from the countryside and increased rural unemployment, rather than the reverse. In Costa Rica and Panama, plantation subsidies were introduced in the express hope that they would help slow down the felling of natural forests. They were so generous that they did the opposite, encouraging landowners to fell natural forests so that they could replace them with plantations in order to claim the subsidy.

Think of major forest traumas around the world—the annual loss of over 7000 square miles of Amazonian forest during the 1980s;

the plunder of primary forests in Indonesia; the frenzied onslaught today on Russia's *taiga*—and they nearly always have one thing in common: subsidies have aided and abetted their destruction. Most of the world's forests are not owned by private individuals or enterprises, but by governments, the majority of whom have treated forests as a resource to plunder, much as they might exploit non-renewable resources such as bauxite or coal. All too often, they have failed to capture the true market value for their timber, and in doing so have transformed what should be a public gain, in the form of royalties and taxes, into private profits for timber companies, or at least into the wages and pensions of bureaucrats. Subsidies have made harvesting profitable in areas where it would not be, were a free, unsubsidized market to exist, and they have encouraged private companies to strip the best timber from forests and treat the rest as expendable waste. Subsidies have encouraged clear-cutting, rather than selective felling. They have led to a dramatic loss of habitat and wildlife and the uprooting of tribal communities. And they have stripped large areas of the planet of a resource that could, and should, provide us with more than just timber.

So what does the future hold for the Tongass? Lindekugel and his colleagues were championing small-scale timber operations, and Caplan agreed that Alaska would never again see a large-scale logging industry dominated by a few operators. "SEACC has done a good job at visioning what they see as a future industry," he said, "but there still needs to be an infrastructure." Small businesses fashioning high-quality wood into bowls and guitars could not do the lumbering and transporting themselves; the Forest Service still had a role to play in making timber available.

What would happen, I asked, if there were no more public money, no more subsidy? Would there still be a timber industry in Alaska? Caplan sat back in his chair and stared thoughtfully out of the window across the town to the huge cruise ships that towered above Juneau's wharves. "I'd love to try it," he said after lengthy considera-tion. "I'd love to help take the national forest system and this region into an approach that said: you've got five years to break even, and we'll fund you on a declining basis for the next five years while you try to sort out the best ways to charge the public for the services they're presently receiving, whether it's timber, or recreation, or minerals, or anything else."

"You think you can make a go of it?" I asked. I was astonished that one of the most influential foresters in the United States should contemplate operating in an unsubsidized world.

"I believe we could, and be very successful," he replied.

"And would that include some timber?"

"It might. I'm not sure." Caplan said that he would charge what the market would bear for recreation, and he would charge hunters and sport fishers, too. Then there were the mineral operations. "I could collect up to 10 percent royalties on mineral development," he suggested cheerfully, adding: "If I could act more like a private land manager, I believe we would be able to return money to the Treasury, and at the same time meet or exceed all existing environmental laws."

While we were on the mine-sweeper, Zencey pointed out that although this was a pleasant way in which to observe the grandeur of the landscape—especially with a drink in one hand, and a canapé in the other—it was not the best way to view the damage done to the forests, because much of the logging had taken place in the rugged interior, out of sight of the coast. Flying south from Juneau, I could see what he meant: from the air the destruction was evident. But in Alaska there were still sizable chunks of pristine forest left intact, whereas you couldn't say the same for much of Washington state or Oregon. The pillage there had been on a scale that is hard to describe, and it struck me as more shocking still that much of this had occurred on public land, managed supposedly on behalf of all Americans. If they really knew what was being done with their money, to land that was in theory theirs, would they stand for it much longer?

I said something along these lines to the gentleman from South Carolina who sat next to me on the way south. He was an engineer, nut-brown from the sun, and silver-haired, and he spoke with a fine Southern accent. This year, as every other year, he had taken his vacation in Alaska and fished all day, every day. The previous year he and his companion had taken home 1000 pounds of salmon and halibut—within a few months, he said, neither he nor his family could bear the sight of fish any longer—but this year the fishing had not been so good. He thought the weather was partly to blame. He said he had no idea that his taxes were being used to subsidize logging in the Tongass, or anywhere else for that matter. How would he feel, I asked, if he had to pay the Forestry Service a fee to fish for salmon? The fish, after all, depended upon the health of the forests for their survival. He thought about this for a while. "I guess I'd rather not," he reflected, "but if paying to fish means that you get good fishing, then I would, and that's a fact."

Alaska had been a good place to start my inquiries, not least because the relationship between subsidies and environmental destruction was plain to see, as was the significant role played by Alaska's politicians in filching from the public purse. There would be times, over the coming months, when I found it far harder to unravel the cause-and-effect of subsidies. But not here. As much as I sympathized with the mayor and his allies, it was impossible to construct a rational argument in favor of the subsidized timber program.

2

CASHING IN
ON COWS?

Wally McRae could have been in his mid-fifties, or he might have been nearer seventy: he had that sort of a face. He had a bushy, sickle-shaped mustache and a complexion the color of an eggplant, and he was elegantly attired in a fancy embroidered waistcoat, cowboy boots, and a bowler hat. Of the 100 or so poets at the 14th Montana Cowboy Poetry Gathering, held in the decorous setting of Lewistown, he was probably the best known. He was certainly the most charismatic, and he treated us to a memorable cameo of cowboy poetry. One of his poems was a send-up of Percy Bysshe Shelley's *Ode to a Skylark*; another hinted, more seriously, at the problems that threaten the rural way of life in the American West.

Once the poetry workshop was over, and shortly before McRae headed back to his cattle ranch near Colestrip, a little way north of the Crow Indian Reservation in east Montana, we fell into conversation and talked about the tough time many ranchers were having. "The future's not bright," said McRae in a slow, sonorous voice that sounded as though it had been matured in malt whisky. "I've got historic ranches all round me. Some of them have been there since we whooped the Indians, and I think two of them will sell up next year. They say that if you will your ranch to your children, you should be prosecuted for child abuse! And that's probably true."

McRae didn't anticipate that he would go out of business in the near future, but life was far from easy. "We run an efficient ranch," he said, "and this year we might make a little money—but that's because we're selling some timber." He thought that the problems for ranchers stemmed not just from too many people producing too much for the existing markets, but from the nature of the modern beef industry. Over four-fifths of all the beef produced in the United States is bought by four packing companies; by acting as a sort of cartel, they have persistently depressed prices.

Wherever I went in the West, I heard stories of hardship among family ranchers. John Paugh, an elderly sheep farmer near Belgrade, reckoned that the last three years had been tougher than the worst years of the 1930s' depression, which he lived through as a child. Since then there had been good times and bad times. "Back in 1946," he recalled, "farmers were getting US$2.50 for a bushel of corn. You can't get that much now." Two years ago wool fetched US$1.40 a pound; now it was down to 40 cents. At the poetry gathering, Jim Hamilton, a rancher from Wyoming, told me that in 1979 he got US$1.10 a pound for his calves; this year he had received 86 cents a pound for the same quality livestock. "And in the meantime," he said, "everything else we need has gone up in price. If you bought a 3-ton pick-up in 1979, it cost around US$10,000. Now a similar truck costs over US$30,000."

McRae was considered something of an iconoclast by his fellow ranchers, not so much because he wrote fine poetry, but because his views were at variance with orthodox agricultural opinion. He was highly critical of the various farming organizations—the local stock-growers' associations, the Farm Bureau, the National Cattlemen's Beef Association—which had failed to make a stand on the ranchers' behalf against the rapacious packing industry. He was also broadly sympathetic to the concerns of the environmental lobby.

This had something to do with the many years he had spent fighting against the coal companies, whose operations had disfigured the land around his ranch. But he was also an intelligent, well-traveled, articulate character who was quite prepared to hold what would seem, to many of his peers, heretical views.

There were no environmentalists at the poetry gathering. Environmentalists do not, as a rule, wear Stetsons, calf-high boots, boot-lace ties, neatly pressed jeans, and huge belt buckles, as most of the men did here. Nor do they walk with a rolling gait that suggests a lifetime spent on a horse. The uniformity of dress among the cowboy poets and the audience was matched by a uniformity of attitude—McRae excepted—towards environmentalists, whom I frequently heard being excoriated, sometimes in the poetry itself, more often in the bar and dining room. As far as many ranchers are concerned, environmentalists are as much of a threat to their livelihood as the ailing markets. They are campaigning against the grazing of livestock on public land, which means half or more of all land in some western states; they are forever putting pressure on federal agencies and state governments to introduce laws and regulations that further restrict what ranchers can do; and they are calling for an end to the predator-control program run by an agency within the United States Department of Agriculture (USDA). "In my opinion," said Jim Hamilton, whose rich, slow drawl reminded me of the actor James Stewart, "the rights of the human are more important than the rights of the three-toed tree toad." Unfortunately, environmentalists disagree. "I go along with the Bible," he said. "Somewhere it says something about man having dominion over all things. Well, I believe that."

I had come to the West to investigate two subsidies that environmentalists claimed were having a damaging impact upon the environment. One—the subject of this chapter—provided for cheap grazing on public land managed by federal agencies. According to many environmental groups, this was leading to the destruction of habitat and the loss of wildlife. The other, the subject of Chapter 5, was the US$10 million annual appropriation that went to the USDA's Wildlife Services. This subsidy was used to kill coyotes, mountain lions, wolves, bears, and other 'varmint'.

Over the past few weeks, during my wanderings around Montana and Wyoming, I had met many ranchers who maintained that there were no subsidies to graze public land. What, I inquired of McRae, was his view? He sighed—he was a great one for sighing—and shook his head. "I guess the enviros are right," he said after a while. "But look at every industry in virtually every country. They're all subsidized. Yeah, so is the livestock industry, but not to the degree that many others are." He added that I would hear a great deal of nonsense talked about some of these issues, both by ranchers and by their environmental opponents, and he suggested that I treat whatever people told me with extreme caution.

I already realized that unearthing the truth was going to be difficult. I had met ranchers who swore that they and their colleagues received not a cent in subsidies, yet organizations such as Friends of the Earth were adamant that the Western livestock industry soaked up over US$500 million a year in a whole gamut of subsidies and tax dodges. I met ranchers such as McRae who said that they had never lost a calf to a coyote; and I met ranchers from the same area who claimed that coyotes killed one in ten of their calves. I met ranchers who believed that they had improved the grasslands, and environmentalists who said that a heavily subsidized livestock industry had all but destroyed the ecology of the West. In the light of all of this, the forestry issue now seemed remarkably simple: large dollops of public money were being channeled toward a small number of large corporations, with the Forest Service acting as money launderer. The cause and effect was obvious, too: taxpayers' money was translated into forest clear-cuts, soil erosion, clogged-up streams, and the loss of wildlife.

The issue of livestock farming and state support is far more complex, for a variety of reasons. Firstly, and most obviously, the potential beneficiaries of federal aid are numbered in the tens of thousands—Montana alone has 13,200 ranchers—and they vary in size from a small number of massive operations running industrial-sized herds of cattle to tiny enterprises with no more than a few dozen animals, which the owners know as intimately as they do their dogs and wives. Secondly, subsidies to livestock farmers—they do exist, contrary to what some ranchers told me—come in many guises, ranging from export subsidies and predator-control programs, to cheap irrigation water, and emergency flood and drought relief. While one rancher may take advantage of, say, the predator-control program, another will benefit from cheap water. Some may tap into a range of subsidies, while others benefit from none at all. The issue is further complicated by the great range of federal and state organizations who play a role in the livestock industry, either as leasers of land or dispensers of grants and loans.

Then there is the issue of environmental damage, which is much harder to discern than it is in the case of forest clear-cuts. "This probably looks fine to you," said George Wuerthner, who took me out to the countryside north of Livingston, Montana, shortly before I went to the poetry gathering. And, indeed, to my eyes it looked lovely: a patchwork of little meadows, spattered with old clapboard barns and farmsteads, gave way to rolling hills, reminding me of the landscapes that figure in the mellower sort of pre-spaghetti western. But to Wuerthner's tutored eye, the landscape was a disaster: the riversides were trampled by cattle and denuded of scrub, the meadows were full of weeds, and there was scarcely a bird to be seen or heard. It had been grazed to hell, as he put it. But overgrazing, I was to discover, was not a universal phenomenon, and it would be unfair to denigrate all ranchers because of the poor husbandry practices of some, just as it would be unreasonable to accuse them all of being "welfare cowboys."

What nobody disputes is that we are seeing a new struggle for the American West, and at its heart is the issue of how public funds, and public lands, are used or misused. The descendants of the men and women who took the land off the Crow, the Cheyenne and the Sioux over a century ago, now find themselves in danger of going under. Their fate, and that of the ranching industry, depends to a considerable extent upon federal policies, and on the way in which public money is allocated. But before we look at the cowboy crisis, and the environmental problems in the West, we should take a brief look at the big picture and establish precisely why, and to what extent, governments support agriculture.

*   *   *

Even if you have taken only a cursory interest in farming matters, you will know that agriculture is the last bastion of the command-and-control economy. With a few notable exceptions—New Zealand is the most obvious—virtually every country in the world provides considerable support for its farmers. Perhaps we shouldn't be too surprised: food is the stuff of life, and it is in every government's best interests to ensure that its people have access to an adequate supply at affordable prices. Telling the citizenry to eat cake when the bread runs out did Marie Antoinette no good—she was guillotined—and governments will do all they can to ensure that farmers keep farming and that food supplies do not run short. In the United States the great exodus of Okies during the 1930s' depression, immortalized in literature by John Steinbeck's *Grapes of Wrath*, is now part of the nation's folk memory. In the United Kingdom (UK), the older generation can still recall the years of food rationing that began with the German U-boat blockade during World War II. Since then the United States and the UK, together with the UK's partners in the European Union (EU), have done everything in their power to encourage agricultural production—generally at the expense of taxpayers and consumers.

Although the United States and the EU have differing approaches to supporting farmers—the EU offers more market-price support, and the United States focuses more on direct payments—the outcome is similar: both shelter their farmers from the free market and encourage oversupply in many commodities. The irony is that this policy has hurt farmers too. Artificially high prices during the mid-1990s led to overproduction; this, combined with the Asian economic crisis and a fall in demand, led to falling commodity prices during the late 1990s. Governments on both sides of the Atlantic reacted to falling farm incomes by offering additional measures to support the producers, placing a further burden on the taxpayer.

In the year 2000, total support to farmers in the 30 mostly Western countries who are members of the Organisation for Economic Co-operation and Development (OECD) was estimated at US$327 billion. Transfers to farmers from consumers amounted to US$174 billion—this is what they would have saved had they been able to buy food at world market prices—while the taxpayers coughed up US$172 billion through price support and other measures. In the EU, total farm subsidies amounted to US$112 billion, in the United States to US$92 billion.

In the year 2000, support to farmers and producers in the OECD, when expressed as a percentage of gross farm receipts, stood at a staggeringly high level of 34 percent. In other words, over one third of what farmers received was a subsidy of some sort. However, this figure masked great variations between countries. In Norway, Iceland, and Japan, subsidies accounted for over 60 percent of farm gate receipts. In the EU, the figure was 38 percent and in the United States 22 percent. Compare this with New Zealand, who has virtually done away with subsidies: there, a mere 1 percent of gross farm income derived from external support measures. In Norway, one of the most protectionist countries in the OECD, the average farmer received over US$30,000 of support. This compared with US$20,803 for the average American farmer, US$16,028 for the average EU farmer and US$338 for the average farmer in New Zealand. Since then, the 2002 Farm Bill has dramatically increased farm spending in the United States.

While it is the developed world that accounts for the bulk of agricultural support, subsidies also play a key role in the agricultural policies of many poorer nations. In many parts of Africa, Asia, and South America, farmers make up a significant proportion of the population. Keeping them happy, or at least pacified, is a political necessity. But the burgeoning cities, frequently beset by poverty, must also be kept fed, so governments tend to keep food prices low to appease the urban poor and compensate farmers by providing cheap and subsidized inputs.

*   *   *

Among livestock farmers in the United States, milk producers receive the lion's share of support. However, as far as environmentalists are concerned, it is the support to the Western livestock producers that really needs to be stanched. Subsidies to dairy farmers may not make much fiscal sense, but they are not perverse, in the sense that they are not, for the most part, environmentally destructive. This, at least, is the way environmentalists see it. In contrast, they say, subsidies that go to Western ranchers certainly are perverse, and among the most significant are those that enable ranchers to graze public land at a cost far below what they would pay for the right to graze private land. This is an issue that has sparked off some bitter conflicts between environmentalists and ranchers, as I learnt when I called on George Wuerthner.

A stringy man with cadaverous features and a high, whiny voice, Wuerthner makes his living as a photographer, but he also describes himself as a grazing activist. More accurately, he should call himself an anti-grazing activist: he is incandescently angry about the damage that the livestock industry has done to the West. "The greatest source of non-point pollution in the West—that's pollution that doesn't come out of a factory pipe—comes from livestock production," he said excitedly over a coffee in his suburban house in Livingston, Montana. "The greatest cause of soil erosion is livestock production. The greatest cause of species endangerment comes from livestock production." And he reeled off a long list of species that had suffered from ranching: grizzly bear, wolf, bison, black-footed ferret, sage grouse, cut-throat trout. Furthermore, said Wuerthner, 97 percent of all of the water used in Montana was taken up by agriculture, much of it to irrigate alfalfa, which was used as winter fodder for cattle. Irrigators received huge subsidies. And then there was what Wuerthner called the avoided costs, which in his view were subsidies too. "I'll give you an example," he said. "A 1600-acre farm near here is on the market for US$10 million. That rancher pays just US$450 in taxes on the property a year. I pay three times more than that for this house, and yet I receive a fraction of the public services which he gets—highway maintenance, water supply, and so on. It's outrageous." And grazing on public land was also subsidized, he added. And for what? To produce meager quantities of beef. Some 96 percent of US beef is reared in the East and South, in states such as Florida and Georgia, where there is plentiful rainfall and good grass. That made sense: in those verdant states you could raise a cow on 2 acres. In parts of the West, you were lucky to get half a dozen cattle on a 640-acre section, and even at that low density they often caused serious damage to the environment.

Once we had finished our coffee we drove around some of the ranches in the parched hills outside Livingston. We inspected trampled streams, weedy fields, eroded gullies, and much else that seemed to confirm Wuerthner's thesis that grazing cattle could have a damaging impact upon the landscape. Strangely, we saw hardly any cattle. Wuerthner thought that they were probably on higher ground; but still, the paucity of cattle told a story: in this dry climate, it didn't need many cattle to make a mess of the place. What's more, said Wuerthner, the cattle that grazed in the West had to be fattened up on grain in feed lots, mostly in the Midwest and the East; and somewhere in the region of 77 million acres of corn were grown in the United States, much of it to feed cattle. Add to this 60 million acres of hay production and you could see that cattle had an extra-ordinary impact upon the American landscape. "I'll put this into perspective for you," said Wuerthner. "All the vegetables consumed in the United States are grown on just 3 million acres of land."

Wuerthner was fighting cattle culture on two fronts: he campaigned against it publicly, and he never ate domestic livestock. When I told him I was running low on fuel, he said, eyeing the fuel gauge nervously, "It wouldn't be a good idea to run out of gas here. I'd hate to have to ask a rancher for help. They might know me." His trenchant views were well known, as he frequently wrote letters to the editors of local newspapers, highlighting such things as overgrazing. He no longer dared to enter local bars alone, because cowboys had hurled glasses at him in the past. He had also received phone calls from ranchers threatening to burn down his house. One rancher even wrote to the local paper to deny Wuerthner's assertion that farming is the greatest source of non-point pollution in the West, and he ended his letter by speculating that the local river would be more polluted still once it ran red with Wuerthner's blood. The following week Wuerthner and his family were leaving to set up a new home in Oregon, and he was clearly relieved: in Oregon, he said, there was none of this redneck aggression toward environmentalists.

The tension between ranchers and environmentalists is just as marked in New Mexico, and shortly before I arrived in this bewilderingly beautiful state a pipe bomb was planted in the Santa Fe offices of Forest Guardians, one of the groups campaigning for an end to livestock grazing on public land. "I try not to think about it," said John Horning as we headed out of Santa Fe one hot September morning, past the sprawl of motels and fast-food joints, many designed in tacky imitation of the adobe Indian pueblos for which the state is famous, then out onto the dry plains toward the Santa Fe River. Fortunately for Horning and his colleagues at Forest Guardians, the bomb did not explode, but it had certainly unnerved them. Who, I asked, had sent it? He said he had a shrewd idea, but was not prepared to reveal who the likely culprits were. He admitted, however, that they made a living from ranching cattle.

A couple of years ago Forest Guardians had outbid a rancher for a grazing allotment along the river. It promptly fenced it off and excluded all cattle. I would be astonished, Horning assured me, by the difference that a rest from grazing makes. We climbed out of his truck and walked a little distance off the road toward the river, which was no more than a foot deep and 5-feet across. "In the arid South-West," said Horning, "streams are the arteries of life, but they've been so clogged with cattle and so overgrazed that they've lost nearly all their cottonwood and willow." On the grazed side of the fence the riverside was almost bare; on the side where cattle had been excluded, a dense thicket of scrub, in places 6-feet high, had grown up over the past two years. Above the scrub, the air was thick with insects.

Horning explained that around four-fifths of all of the songbirds found in the South-West depended for their survival upon riverside vegetation, most of which had been lost in New Mexico, primarily as a result of cattle grazing. According to Horning, grazing had done more damage to wildlife, and been responsible for a greater loss of habitat and species, than logging, mining, urban development, or any other activity in the West. However, environmentalists who challenged livestock grazing in states such as New Mexico, found it difficult to convince the public that there was a problem. "We're up against the myth of the cowboy," said Horning, adding that it was like campaigning against apple pie and motherhood. This frustrated him deeply.

For people like Horning the myth of the cowboy is all about deception. Cowboys, in his view, have fooled the rest of the world into believing that they are upstanding, morally superior frontiersmen who rely not on government handouts, but on their own wits and ingenuity. Imagine James Stewart in one of his more sentimental roles, but with a good punch all the same, and you get the picture. Many cowboys contend that this is not a myth at all. This is how they, or most of them, really are. They are people—and this is one of their favorite aphorisms—who mean what they say, and say what they mean. They are people who look after their neighbors, who you can rely on in times of trouble. They are God-fearing, yet tough; uncompromising, yet fair.

Whatever view you take—I personally found much to admire in cowboy culture—there is no denying the remarkable significance of the American cowboy. As Hugh Brogan points out in his *History of the United States of America*, "The legend of the West, the Matter of America, is that country's greatest gift to the imagination of the world, and a historian can neglect imagination only at excessive cost." Once he has summoned up the legend—James Stewart riding into town; smoke rising off the mesa; Apaches preparing to attack —Brogan reminds us that legend, unchecked, is the greatest enemy of historical truth. The truth, of course, is that the history of the West was primarily one of conquest, of the vanquishing of the native Indian tribes to make way for settlement, mostly for people of European descent. Not that you see or hear much in the West that explicitly acknowledges this. The rodeos, the fine museums in places such as Cody, a thousand Hollywood films, and many a trashy cowboy novel, have little to say about the treachery which, in Brogan's view, "was a principal theme in the whites' treatment of the red man."

However, this was the past, and in assessing whether, and to what degree, today's ranchers are taking money from the public purse, we should neither burden them with the sins of their fathers, nor be seduced into idolizing them.

After I left Horning, I made my way to Silver City, a mining town in south-west New Mexico whose immediate surroundings are pocked with the vast craters created by the Phelps Dodge copper operations. It is a tough little place, and environmental activists are few and beleaguered. Nevertheless, Mike Sauber of Gila Watch told me that he had been in Silver City so long, and had become such a part of the commercial furniture—he ran a bike shop—that even his most implacable enemies accepted that he was here to stay. However, his partner in Gila Watch, Susan Schock, had been threatened by ranchers so often that she had decided to leave the area to bring up her daughter in a safer environment.

During my time in the United States, I met several courageous environmental activists whom I found it difficult to warm to. A truculence in their manner, a stridency of rhetoric, and a lack of good humor often meant that though I might sometimes sympathize with their cause, I preferred the company of the people—cowboys, loggers, miners—they so despised. Sauber was not like this. Courageous, yes; trenchant in his views, certainly. But he was a warm and personable individual, too.

I called on Sauber, a lean, ginger-bearded man, as he was locking up his bike shop and he suggested we talk at his home. As we headed up the quiet streets with their modest clapboard houses, he became increasingly animated: mention the issue of grazing to Sauber and you can sense his blood pressure rising. The story he told me may not be typical for the whole of the American West; but it highlights the way in which certain factions within the ranching industry abuse public lands, with the blessing of local politicians and many farming organizations. Subsidies are at the heart of this story.

Sauber and Schock set up Gila Watch in 1991 when they heard that the Forest Service intended to build 33 new stock tanks to provide drinking water for cattle in the Gila and Aldo Leopold Wilderness areas. Gila Watch challenged the Forest Service in the courts and won; the plans were deemed illegal under the Wilderness Act. But why 33 tanks? And why was the Forest Service keen to help graziers in this way? Sauber and Schock began to investigate, and they soon discovered that the Forest Service was being put under pressure not just by the intended beneficiary of the tanks— the holder of a permit to graze the 145,000-acre Diamond Bar allotment—but by a Texan bank.

When the former permit holder for the Diamond Bar allotment was declared bankrupt, the Forest Service decided to reduce the number of cattle that could be grazed on the allotment from 1188 to 866. The aim was to reduce overgrazing of the forests. The Forest

Service notified the bank that held the mortgage on the grazing permit. Although such permits are held as "revocable privileges", and do not, in principle, have a pecuniary value, they do not evaporate when a property goes bankrupt; rather, they are held by the Forest Service for the bank until it sells the base property with the grazing permit attached. In this case, ownership of 115 acres of private land gave the permit holder the right to graze 145,000 acres of public land. The bank calculated, correctly, that a permit for fewer cattle would diminish the value of the base property.

"They said they were going to play hardball with the Forest Service, unless it got the numbers back up," recalled Sauber, "and they suggested that the Forest Service should contact the Range Improvement Task Force at New Mexico State University." The Forest Service did this, and the task force suggested that the addition of 33 tanks to areas that previously had been deemed unsuitable for grazing, due to a lack of water, would enable the Forest Service to increase the stocking density. "And surprise, surprise," exclaimed Sauber, "by adding formerly unsuitable land, and by math error, it brought the number back up to 1188!"

It would be wearisome to provide a blow-by-blow account of what happened subsequently. Suffice it to say that the Forest Service did all it could, at first, to keep the permit holder, Kit Laney, happy. "The Forest Service bent over backwards to appease him," said Sauber. "But when they said he had to reduce cattle numbers, he called up the governor, a US senator, and a congressman, and they all came in to bat for him. He even told the Forest Service, 'If you try and move my cattle off, you're going to be met by 100 people with guns.'" Eventually, after various court cases (in one, Gila Watch sued the Forest Service for allowing Laney to damage a wilderness area), Laney was forced to remove his cattle, having—in Sauber's words—"profited from the illegal destruction of the wilderness for a two-year period." Now that the cattle have gone, the land is slowly recovering.

Many months after I left New Mexico, I heard from Sauber again. "PS," he added at the foot of his letter, "Laney illegally ran 200 cattle down the Gila River to set up his new cattle operation on private land. Believe it or not, he also re-applied for his old permit on the Diamond Bar. We will sue if he gets it."

Environmentalists such as Sauber and Horning contend that much of the damage done to nature is attributable to subsidized grazing. By failing to charge the true market value for grazing permits, federal agencies are encouraging ranchers to overstock and overgraze the land, and to run cattle on marginal and ecologically fragile areas. And Americans are loosing out twice: not only is the environment suffering, the government is being deprived of revenue.

The main agencies involved are the Bureau of Land Management (BLM) and the Forest Service. The BLM manages around 165 million acres, and the Forest Service has 50 million acres that it considers suitable for livestock grazing. Both charge a monthly fee for grazing, and under existing legislation this is determined by a formula that takes into account, among other things, the state of the livestock market. The minimum charge per animal unit month—the equivalent to a cow and a calf, or five sheep, grazing for a month—is set at US$1.35: "Less than it costs to feed a hamster," as Horning put it. The same charge applies to all graziers of public land, regardless of rainfall, productivity, access to markets, and so forth.

Environmentalists point out that federal grazing fees are far lower than fees to graze on private land, which are generally at least four or five times higher, the implication being that the difference between the two is a subsidy. All those I met insisted that there was absolutely no disputing the fact that there was a grazing subsidy. For one thing, holding a federal lease increased the value of the associated ranch, or base property, and this suggested that federal grazing fees were set below the true market value. For another, during the 1980s, federal grazing fees fell, while fees for grazing on private land rose: this also suggested that grazing fees were subsidized.

So much for the argument for the prosecution. What about the defense?

You only have to read the agricultural press to realize what a big issue this is, and how incensed many ranchers are by accusations that they are scrounging off the public. This, after all, is not how James Stewart and his ilk behaved. In one of the newspapers I came across, the New Mexico Cattle Growers' Association was soliciting funds to fight against the mendacious propaganda of the Serbs, whom it accused of ethnic cleansing and cultural genocide. The Serbs—selfish environmental radical bigots—were apparently financed by "wealthy outsiders AND the federal government who pay their legal fees and provide YOUR tax dollars through grants and buffalo price supports." The Serbs, according to the advertisement, wanted to cleanse public land—or federal land, as ranchers prefer to call it—of ranchers.

The belief that environmentalists and the federal authorities are in some ways colluding in a war against ranchers is widely held. It is particularly well articulated by People for the USA, a grassroots pressure group who has brought together ranchers, miners, loggers, and others who believe that federal policy is threatening their rights to use and exploit federal lands. At the time of my visit, People for the USA's representative in New Mexico was Betty Hyatt, and she invited me to visit her when I was passing through Deming, an unappealing sprawl of a town on a flat, dusty plain not far from the Mexican border. Having read some of the uncompromising material produced by People for the USA, I expected to find a tough, leathery-skinned, broad-bottomed female, possibly wielding a wood ax, when I arrived at the Hyatt ranch after a 7-mile drive along a dirt track. Instead, I was greeted at the doors of a modern one-story house by a willowy widow with elegant manners, a gracious way of speaking and plenty of style. Imagine Fay Dunaway in *Bonnie and Clyde*, then add 50 years. Betty Hyatt knew that I had already visited Horning and various other "environmental extremists", and she had gone to much trouble to gather together a group of people who would provide me with a vigorous defense of the ranching way of life.

One of these was her son, Leadrue Hyatt, who, with his brother, ran 1500 head of cattle over some 60 sections (a section is 640 acres) of land, a good chunk of which was federal land. Hyatt was a small man, with a fractious and disenchanted air about him, though this may have had something to do with the nature of our conversation that afternoon. He immediately made it clear, in a querulous voice, that he detested "bunny huggers" and "tree squeezers". The next person to arrive was Nick Ashcroft, an economist who worked for the Range Improvement Task Force at New Mexico State University. This is the organization who advised the Forest Service to increase the number of water tanks on the Diamond Bar grazing allotment. A stocky character with a droopy mustache, watery eyes, and a shy, hesitant way of speaking, he allowed the Hyatts, and the last visitor to arrive, Frank DuBois, to do most of the talking. DuBois, the secretary of agriculture for New Mexico, was a man of great charm and bore his infirmity—he had multiple sclerosis—with fortitude. He looked as though he was in his mid-fifties, and he was smartly dressed with a boot-lace tie and a voluminous black hat. He explained, as he lowered himself gingerly into a chair, that his great-grandfather had homesteaded land in New Mexico, but his grandfather had lost it during the Depression. His family still ranched 16 sections, 11 of which were owned by the federal government; but the authorities only allowed them to graze 94 head of cattle. "So I had to seek employment elsewhere," wheezed DuBois, "and that's why I'm a bureaucrat."

Even the medium-sized family ranchers such as the Hyatts were finding it hard to make a living, said DuBois. "What's killing these people now," he said, "is not so much the state of the market, it's the federal government, and federal government policies and the impact they're having."

This was the dominant theme of our conversation, which continued, over iced lemonade and homemade cookies, throughout the afternoon. DuBois explained that there were now over 20 kinds of federal administrative designation that limited the ranchers' ability to manage the federal lands on which they had permits to graze, and which consequently limited their ability to make a profit. Much of the Hyatt grazing allotment, for example, was a Wilderness Study Area. Prior to its designation in 1976, the Hyatts could manage it as they saw fit. Now they couldn't lay pipelines, or use mechanical equipment or off-road vehicles in the designated area. This meant that they could not run it effectively, said Hyatt. The family had wanted to put in pipelines and stock tanks in order to disperse the cattle, but the environmentalists had objected and the federal managing agents had refused the request. As a result, the cattle were now concentrated, and this did far more damage to the environment.

When I raised the issue of subsidies, Hyatt let out a harsh laugh and said: "It makes me guffaw to hear the enviros talk about subsidies, when they're subsidized to hell themselves." In his view, environmental organizations were creaming funds off the government, for one thing and another. But wasn't his grazing subsidized, I asked? No, sir, it was not. It was true that the grazing fees were much lower than the fees to graze private land, but then the costs were so much higher. Private landlords charged US$6 or more per animal unit month and generally provided fencing, water, and other facilities. In contrast, permit holders on federal land had to pay to install fences, water tanks, and the like. In addition, the labor costs and feeding costs tended to be much higher on federal lands. So when all of these factors were taken into account, it was clear that the grazing fees were not subsidized. Ashcroft agreed and gave me an academic paper that suggested that grazing on federal land was not the great deal it was made out to be by the environmentalists.

DuBois was firmly against all subsidies. "We should abolish the lot," he said. "Then the good ranchers and farmers will survive, and the marginal ones will go to the wall." He said that he thought the disaster assistance programs which provided flood and drought-relief subsidies were "bullshit", although Leadrue Hyatt seemed to approve of them. The United States is always giving aid to poor countries when they have earthquakes, so why, he reasoned, shouldn't farmers be given help when they suffer from natural disasters here? There was, however, total agreement among them about what the federal government ought to do. "They should privatize all land," said DuBois. "They should give ranchers the first right of refusal, as they're the ones who've paid for the improvements on the land. I believe the most efficient allocater of resources is the market."

And what would happen if public lands were not privatized, I asked? "I hate to say this," replied DuBois, "but if someone asked me if the Hyatts will be here in 25 years' time, I'd say no. If you look at current trends, the federal authorities are moving livestock off public lands. If there's going to be ranching here, then the federal land will have to be privatized."

I left the Hyatt ranch late in the afternoon and drove around the countryside near Deming. It was mostly flat, and everywhere arid and treeless. Buzzards stared down at the coarse grass and sage brush from telegraph poles, and I passed a coyote which was trotting nonchalantly down a dirt track. There were Herefords and other breeds of cattle, but they were few in number and dwarfed by the vastness of the plains and the great dome of blue sky, which rapidly darkened as the sun set. That night I found a cheap motel on the edge of Deming and went in search of food and drink. Restaurants served food near the motel, but they were alcohol free, and the bars had no food; so I ended up walking 2 miles or more into what passed for a town center. I walked by numerous trailer parks, some with permanent local residents, others with elderly couples who had come south to eke out their pensions, and eventually I arrived at a Chinese restaurant and ordered chop suey and cold beer. While I ate, I reflected upon what I had seen over the past few weeks and how at variance much of it was from what I had anticipated.

I had been told by the environmentalists in Washington, DC, who had urged me to come out to the West that this grazing issue was relatively simple: there was a clear link between subsidies and environmental damage. I don't think it's so simple, and I will complicate matters further by telling you about some other ranchers whom I met during my travels. So far I have dwelt mostly on the two extremes: on the campaigners who deplore cattle culture and believe cattle are devastating the West, and on the ranchers who refuse to accept that they ever do any wrong, and who believe they should be left to do as they please. "So you've seen the crazies from both ends of the spectrum," drawled a BLM employee when I told him whom I had seen. Perhaps, but I had also spent time with many others who fell into neither camp.

There was Chum Howe, for example, who ran 90 head of Highland cattle on an exquisitely beautiful patch of land in the Gallatin Valley to the north of Bozeman, Montana. If you've seen Robert Redford's *A River Runs Through It*, then you have seen Howe's ranch: that is where much of it was filmed. I called in on him during hay time, and his son was cutting grass in the patchwork of fields that lay beneath a great wedge of well-wooded mountainside. We took some sandwiches out to the son, then walked into the forest to search for cattle. Howe explained that if ranchers were to survive—and most his size were in trouble—they had to be ingenious. Rather than producing conventional beef and selling it to the packers, who would then fatten them on corn and steroids in feed lots, Howe was rearing his stock organically on grass and selling to the health food stores around Bozeman. By tapping into a specialized market he was getting four times more for his cattle than ranchers selling "any brand" beef to the packers. He disapproved of all subsidies, and he said he knew other ranchers who were not beholden to the American taxpayer. For years after he took on this ranch, the US Department of Agriculture had sent him forms inviting him to apply for crop subsidies. It was as though the government wanted to throw money at him, he explained, and he was clearly shocked by it.

Then there was John Finlay, whose Scottish grandfather had homesteaded a wild valley—still known as Scottish valley—some 7000 feet up in the great rolling hill country of central Wyoming. A tall, thin man with a diffident manner, a droopy mustache, a weathered complexion, and a florid dress sense, Finlay had just 30 head of cattle, and he supplemented what he made from these by painting wildlife, which he did with great proficiency. I went to see him as I had heard that he once had a wolf on his land. The previous year, he explained, a lone wild wolf—as opposed to one of the reintroduced wolves that were now spilling out of Yellowstone National Park—came to play with his dog outside his kitchen window. The state authorities got wind of this, and the Department of Fish and Game shot the wolf—they claimed it might be a cross between a wolf and a dog—and he was furious with them for having done so. Clearly he loved wildlife. When I asked him what subsidies he received, he said: "Up here we've never received a cent." In fact, his family had once applied for a grant to install water supply equipment, but the application was refused. He grazed his cattle on public land high up above his property, but he said that there was no subsidy involved. By the time you factored in the cost of travel, fencing, and so forth, he was paying as much as he would were he to graze on private land. At present, he was carrying out a survey of the vegetation to establish whether or not his and his neighbors' cattle were causing any problems. Later I rode up to the area on a sure-footed quarter horse. We saw black bear, pronghorn, and coyote, as well as a few cattle and some majestic forest. Clearly, Finlay was a good steward of the land, and I met many others like him.

The truth of the matter is that you see private land that has been chewed over and heavily eroded, where the streams are denuded of scrub and there is no sign of wildlife, and you see public lands like that, too. But there is also plenty of land, both public and private, where ranchers take care not to damage the environment. Nevertheless, surveys suggest that although matters might not be as dire as environmentalists claim, public lands are often in poor condition. According to one government report on federal range lands, 37 percent was in good or excellent condition, 15 percent was in poor condition, and 37 percent was in fair condition. Put another way, this means that 60 percent of federal grazing allotments were in less than satisfactory condition. On the other hand, everyone agrees that the range lands are in far better shape than they were during the early decades of the 20th century, when there were no controls over grazing.

As good a "quasi-official" view as you will get on whether grazing is subsidized or not comes from a 1992 report by the Government Operations Committee. The committee calculated that the federal government had lost US$1.18 billion since 1985 by not pricing grazing fees at their true market value. This represented a loss to the US Treasury of around US$170 million a year. According to another government report, there is a further subsidy in the sense that half of the value of the fees paid by ranchers to the BLM is transferred into a Range Betterment Fund, most of which has been used to fund activities of direct benefit to the graziers. Both the BLM and the Forest Service pay a proportion of their fees to the states, and these are then distributed to the counties where the grazing takes place. Ranching organizations make much of the fact that these fees are used to finance public services. True, they are; but a good portion is channeled into activities that support the ranchers too. These funds have frequently been used to finance livestock organizations, and on one occasion to sue the federal government over a grazing dispute. This is equivalent to a mugger borrowing your hammer, which he then uses to knock you senseless.

You could certainly pick holes in the 1992 report, for all the reasons highlighted by Hyatt and Finlay. However, there is no disputing the loss to the Treasury. The latest *Green Scissors* analysis of the grazing subsidy points out that in 1998 the federal grazing program generated US$22 million, while it cost US$116 million to manage and administer. Put another way, ranchers were paying a mere US$1.35 for each animal unit month to the federal agencies, while costs for the federal agencies amounted to over US$6 for each animal unit month. The loser here is the taxpayer, who has to foot the US$94 million a year bill to cover the deficit. However, does culpability lie with the ranchers, for not paying enough, or the federal agencies, for managing the lands at such high cost?

Organizations such as the Political Economy Research Center (PERC), whom I called upon when I was in Montana, point the finger of blame firmly at the federal agencies. "We're libertarians," explained Bishop Grewell, who with his short hair, fresh complexion, and baseball cap looked as though he had just stepped off a college football field. "We think federal government usually causes more environmental problems than it fixes. When public lands are managed by bureaucrats, they tend to suffer. Bureaucrats simply don't have the same long-term incentive to manage land properly, as private owners do. And they tend to come up with general plans for very diverse lands, so they're often not appropriate." Furthermore, said Grewell, the bureaucrats themselves were major beneficiaries of the subsidy system.

One of Grewell's colleagues, Holly Lippke Fretwell, had recently investigated the profitability of public lands, and in particular the 20 percent of America controlled by the Forest Service and the Bureau of Land Management. According to these agencies' own figures, they managed to lose US$290 million on timber, US$66 million on grazing, and US$355 million on recreation each year between 1994 and 1996. While the federal agencies were squandering public resources, Fretwell found that state land-managing agencies were generating considerable revenues from the land in their control. Take, for example, grazing on public lands. Every dollar spent on range land management by the Forest Service and BLM generated just 25 cents. State trust lands, in contrast, earned over US$7 for every dollar spent.

When I put these figures to employees of the BLM in Billings, Montana, they said something along the lines of: yes, we can believe that; we employ a lot more people than the state does. Indeed, the federal agencies employ four times more people for every million cattle grazed than the states. The states have to be efficient in their use of the land as the revenues are so important: they are used, among other things, to fund the public school system. Unfortunately, there are no such incentives for federal agencies, which rely on Congress for their budgets, and return their revenues to the Treasury. While state employees generate annual revenues of around US$425,000 each, the Forest Service employees earn the Treasury less than US$24,000. In the view of the libertarians, this means that when we talk of subsidies to public land users—whether they are livestock farmers, mineral operators, timber companies, or hikers —we must count the bureaucrats as beneficiaries, too.

So where does this leave us? I would suggest that in an attempt to simplify a highly complex issue, we should deal with the issue of environmental damage caused by grazing separately from the issue of subsidies, even though the two may at times be related. It obviously makes no sense to allow private ranchers to damage a public resource. If there is clear evidence of overgrazing and habitat destruction, then the grazing should cease, and the land would be better off left to nature's devices and desires, or put to some other commercial purpose that does not damage it. Take, for example, the case of the Diamond Bar allotment. Here was a private individual who had the right to graze—let us take one of the middle figures— 650 cattle on 145,000 acres of federal land. This was the equivalent of one cow every 223 acres, and even at this low density, the livestock were frequently malnourished. They were also causing serious damage to the forest. In Georgia and Florida, where much of America's beef is reared, it takes less than 2 acres to support a cow. So where is the sense in subjecting public lands in places such as New Mexico to an activity of negligible economic significance, at considerable cost to the environment? There isn't any.

In those parts of the West where it is possible to graze cattle on public land without adversely affecting the habitat—in other words, where cattle ranching is sustainable—then it would make sense to ensure that there is no burden on the taxpayer. At present there is, and two obvious remedies exist. Firstly, there should be a system of competitive bidding for grazing allotments: this would establish their true market value, and the value—unlike present grazing fees— would reflect the many variables that influence the profitability of grazing on federal land. Secondly, the federal agencies should be obliged to at least cover their costs, and preferably to make a profit. At present, they are failing miserably, and they are receiving a direct appropriation from the taxpayer.

There is another remedy that will not find favor with the free-market, pro-privatize lobby, or with the anti-cattle environmentalists. Many months after I left the West, I was chewing the cud with Celia Boddington, an acidly witty Englishwoman who runs the BLM's public affairs department in Washington, DC. She was incensed by the comparison that the Political Economy Research Center had made between state land and federal land grazing. "The state trust lands are often the best lands," she said indignantly. "The states chose the best land because they knew they would have to raise revenue for public services. Our land is often the worst—it was what the homesteaders didn't want." Furthermore, the states were able to charge royalties for hard-rock mineral operations on their land, whereas the federal agencies could not.

As it happened, the BLM raised more money than it spent, and was a major revenue earner, thanks largely to the royalties it charges for oil and gas exploitation. However, Congress had given it a role which the libertarians at PERC and the Cato Institute chose to ignore. The BLM was supposed to manage the land for a whole range of activities, from livestock grazing to public recreation and wildlife conservation. Privatizing the land, as PERC and the ranchers at People for the USA urged, would be a disaster, according to Boddington. It would simply lead to the high-grading of public lands: "The choice bits would be bought up, and the whole process would be driven by money." Oil companies would cherry-pick what they wanted, mineral corporations would exploit what they wanted, and nature conservationists would take what they wanted. The market would be flooded with property, and private property prices would collapse.

So what, I asked Boddington, would she do about the grazing fee issue? "Well," she said, "I am speaking for myself, not for the BLM. I think grazing fees are a complete red herring. If we are really concerned about the future of public land in the West, we should forget the fees. It costs more to administer and collect than we get. Let's use what resources we have to manage the land better. Let's stop vilifying ranchers and help them to look after the land." In the sense that ranchers would be getting something—grazing—for nothing, this would obviously be a subsidy, but it would cost the public less than the present system. The Boddington plan struck me as pragmatic. Politically, neither the right-wing "let's-privatize-it-all" lobby, nor the "let's-stop-all-grazing-on-public-land" lobby has a hope of getting what it wants. Grazing on public land will continue—the myth of the cowboy will make sure of that—and it would make sense for ranchers and conservation-minded federal agencies to work together to create and maintain a healthier landscape than we have at present.

*   *   *

I had headed out West to investigate the grazing subsidies largely because environmentalists considered them to be a great scandal. Now I realized that they were wrong: there were other far more wasteful, and equally destructive, livestock subsidies, some of which they ignored. Most obviously, there was the dairy program, about which they have remained strangely silent. The argument that the dairy industry is environmentally benign simply doesn't stack up. Intensive dairy units produce vast quantities of slurry, and the animals are fed on grain and grass whose cultivation necessitates the use of artificial fertilizers and pesticides, and brings land under the plow which might otherwise have remained in a more natural state. The dairy program also fleeces the US taxpayer in a grand way, accounting, in 2000, for US$11 billion of state support. This compares with US$1.3 billion of price support and other payments to the US beef sector. While state support to beef farmers accounts for 4 percent of their gross earnings, support to dairy farmers accounts for around half.

So why have the greens made so little fuss about dairy subsidies, or, for that matter, corn and wheat subsidies, when between them they accounted for 52 percent of direct farm spending in 2000, and when the impact on the environment of these forms of agriculture is plainly considerable? Jerry Taylor and Steve Slivinski of the Cato Institute think that they know why. Responding to the 2001 *Green Scissors* report, they noted that it denounced the sugar, tobacco, mohair, and cotton programs, all of which largely benefited farmers in Republican districts. "Fine, we agree," they wrote. "But where's the milk program? Are milk subsidies somehow less obnoxious than the rest?" The answer is no. The implication is that the greens are all too eager to decry subsidies that benefit Republicans, but strangely shy of challenging Democratic interests. There is undoubtedly a case to answer.

Not that Western livestock graziers are living entirely on their own wits and resources. They aren't, but grazing subsidies—if you accept that that is what they are—are dwarfed by all manner of other subsidies to the livestock industry. Some are implicit, and involve undercharging for federally managed resources, such as water; this is a subject that I discuss later. Some, on the other hand, are direct, and involve payments to ranchers and farmers.

One livestock program that has outraged both the fiscal conservatives of the Cato Institute and the *Green Scissors* campaigners is the mohair subsidy, which was first established in 1954. During those days, when army uniforms were made from wool, the government was concerned that there would be a dearth of material if the United States went to war again. Loans, price supports, and other measures encouraged farmers to produce mohair from goats and wool from sheep. Times changed, but government policy didn't. By the early 1990s, there was a worldwide glut of wool—and, in any case, synthetic fibers had usurped natural fibers in the manufacture of many articles of clothing. Yet millions of dollars of subsidies still found their way to some 100,000 producers in the United States, costing the taxpayer some US$200 million a year. In 1994, Congress at last saw sense and voted to phase out the program. Yet five years later, the subsidy rose, phoenix-like, from the ashes: mohair was once again made eligible for interest-free government loans, and these are now enshrined in the 2002 Farm Bill, as are direct payments to mohair producers. Besides hurting the taxpayer, the subsidy also hurts the environment. Grazing goats compete with native species such as white-tailed deer for water and grass, and goats are a menace to ground-nesting birds.

Earlier in this chapter I quoted the levels of support received by the *average* full-time farmer in various countries. Of course, there is no such thing as an average farmer, and in a system that generally links levels of support to the scale of production and the size of enterprise, it is the big producers who corner most of the benefits. In the United States, almost one third of all agricultural support goes to just 2 percent of the farmers; and over four-fifths goes to the top 30 percent. A similar story can be told for the countries of the EU. The larger, better-off farmers get the bulk of public support, although there have been some explicit attempts to keep small farmers in business in marginal areas by providing a variety of subsidies.

Leadrue Hyatt wanted to know if I had read the novels of the vet James Herriott. These are set in the Yorkshire Dales, in the north of England. Not only had I read some of them, I was able to tell him that I knew farmers who had used Herriott's services as a vet, and I had spent a good deal of my youth in the countryside about which he wrote. The Dales possess one of the loveliest of landscapes in Europe. Small cornfields and hay meadows line sinuous, rushing rivers such as the Yore, the Wharf and the Swale, and undulating farmland, punctuated by small woodlands and handsome stone villages, rises up to the distant moors, which are a spectacular purple when the heather is in flower in late summer. Unlike much of what you see in the American West, the landscape of the Dales—and the landscape throughout much of Europe—is manmade. Thousands of years ago our ancestors carved fields out of the dense forest, and virtually everything you see in the Dales today—the hawthorn hedges, the limestone walls, the heather moors, the flower-rich meadows—is there to support, feed, or restrain livestock. In the valleys the grasslands are grazed by dairy cattle; on the higher, less fertile ground, the land is given over to beef and sheep.

Were the farmers to cease farming, then most of this would revert to scrub. And if it were not for subsidies, that is exactly what would happen on the higher ground, where livestock farmers would be unable to survive. Indeed, even with subsidies, many are being forced to pack their bags and sell their livestock. Over the past few years, the prices that farmers received for their sheep and beef plummeted, for a variety of reasons. The collapse of the Russian rouble meant that there was no longer a market for sheep skins. Mad cow disease—or bovine spongiform encephalitis (BSE)—meant that farmers could not sell their beef to continental Europe. Ludicrous EU regulations led to the closure of small abattoirs, and drove up farmers' costs. A strong UK pound has meant that British produce has been overvalued compared to that of other EU countries. And foot and mouth disease has inflicted huge damage to certain regions. But subsidies have also been to blame: they have often encouraged overproduction.

Take the sheep sector. EU sheep farmers receive around 3 billion Euros a year in subsidies. These include headage payments: the more stock you have, the larger your subsidy check. Because farmers receive an annual check for each breeding ewe, there is no incentive to slaughter them. So even though sheep prices in the UK in 1998 were around half of what they were the previous year, the national flock still grew some 10 percent over the next year, thus pushing prices down even further. It would be wrong to blame overproduction entirely on subsidies: falling farm gate prices have also encouraged farmers to increase stocking densities. However, it is fair to say that subsidies linked to production encourage overproduction; overproduction leads to lower prices; lower prices lead to debt and bankruptcy; and the descent into indebtedness often forces farmers to squeeze more from the land in a desperate effort to survive.

*   *   *

It came as something of a surprise to find that many ranchers were openly hostile to subsidies. Wally McRae said they should be done away. Chum Howe, a staunch Republican, was appalled by them. "We should abolish the lot," echoed Frank DuBois emphatically. Many others said the same, but nearly all added that it would only be fair if those who made a living in the same way had to live by the same rules.

In fact, there is one word that is never far from the lips of farmers and ranchers, and those who represent them: unfair. Sheep farmers in Montana complained to me that New Zealand sheep farmers, whose produce was pushing local lamb off the grocery shelves in Montana, were at an unfair advantage because they had no predators to worry about. But then the New Zealanders could counter that they get none of the subsidies that benefit many ranchers in Montana. In New Mexico, I talked to ranchers who said that they had difficulty in competing with Mexican beef and Mexican chilies: the former was produced with cheap labor; the latter was dowsed with pesticides banned in the United States. It was unfair. On my return to England I visited farming friends in North Yorkshire. One of them had paid twice as much to have his sheep sheared as he received for his wool. This, he explained, was because there was a glut of wool, not least because Australia was producing so much, and Australian ranchers were not subject to the same welfare standards as UK farmers. It was unfair.

Virtually every farmer and rancher I met told me that what they needed was "a level playing field". They wanted trade to be fair. Well, I would suggest that the best way to achieve this is not by protectionist trade policies, or by robbing the taxpayer and consumer to pay the farmer. Rather, we should seek to dismantle as much as we can of the subsidy system. At the very least, we need to abolish all subsidies that are tied to production. These often favor the big and powerful, to the detriment of small family farmers, and frequently they encourage farmers to rear and grow too much food. The libertarians argue that all we need to do is embrace free trade. But free trade is not necessarily fair trade. Without certain checks, it will simply mean that the most voracious exploiters of capital, land, and labor, and those who are willing to sacrifice the welfare of livestock on the altar of profit, will produce the cheapest food, and win out at the expense of decent husbandry.

The champions of free trade in its purest form would argue that guaranteed prices, tariffs, import and export quotas, and all the other complex paraphernalia of support should be abolished. As evidence they can point to New Zealand, which provides a telling example of the benefits that flow from abolishing subsidies. In 1984 virtually all farm support ceased. Farm output initially declined, but later recovered; and there are more farms now than there were in 1984. The abolition of subsidies also brought an end to the clearance of marginal land. It has not been an easy process, but it has forced farmers to become more innovative and to compete more efficiently in the marketplace.

However, New Zealand has a system of law that ensures that farmers must pay a living wage to their employees; and that they do not mistreat their livestock, pollute their rivers, or threaten endangered species (although, admittedly, many are endangered as a result of past farming practices). The same cannot be said for many other nations, where there is the very real possibility that free trade, in its purest form, would reward those who plunder the environment, and abuse both humans and livestock. This would be unacceptable, and in Chapter 10 I will try to steer a course through this minefield.

There might—and I stress *might*—be a case for individual nations to decide that certain sectors of the agricultural industry are worth preserving, even if they are, from a macro-economist's point of view, uneconomic. Perhaps the survival of individuals in these sectors benefits society as a whole because they look after the land and preserve landscapes that would otherwise disappear, or perhaps they help to preserve rural communities. This does not mean that they should be awarded subsidies related to their productivity, or to the size of their holdings; or that they should be protected in the marketplace by import tariffs and quotas. They should be treated, instead, as people who are doing a job on our behalf, and on behalf of those among whom they live. But this is contentious stuff, and I shall return to it later.

CHAPTER 3

SWEET CHARITY

As Representative Dan Miller ushered me into his spacious office on Washington's Capitol Hill, I reflected that he was the embodiment of Republican respectability. Lean, sun-tanned and silver-haired, with a smile a toothpaste salesman would die for, he was the type, I imagined, who gave generously of his time and money to the local church, who worried about the poor without wanting to throw money at them, and who probably was not much interested in the fate of endangered species, or the music of Eminem. His clean-cut young aides looked the part, too. They exuded moral rectitude, and the cuttings pasted on their filing cabinets mocked the lurid goings-on at the White House—President Clinton's affair with Monica Lewinsky was still front-page news. So I was somewhat taken aback when Miller asked, as soon as we settled into comfortable leather armchairs, "Have you read a book called *Striptease*?" I said I hadn't, and he suggested that I get hold of this Carl Hiasson blockbuster as soon as I left.

The opening scene, in a Fort Lauderdale strip club, finds Representative Dilbeck—like Miller, a Republican from Florida—worse for wear. He watches an exuberant gang of young men spray champagne over the dancers; then, outraged by the sight of one of them clinging to a dancer's waist, he batters him senseless with a champagne bottle. One of his first pronouncements is: "I love naked women, I truly do." After some ten pages I realized why Miller had pointed me in the direction of this entertaining book. It was, among other things, about subsidies—specifically, the sugar subsidies in Florida—and the way in which big business buys political favors. As Hiasson explains:

> Every few years, the Congress of the United States of America voted generous price support for a handful of agricultural millionaires in the great state of Florida. The crop that made them millionaires was sugar, the price of which was grossly inflated and guaranteed by the US government. This brazen act of plunder accomplished two things: it kept American growers very wealthy, and it undercut the struggling economies of poor Caribbean nations, which couldn't sell their own bounties of cane to the United States at even half the bogus rate.

And to make this happen, "Big Sugar" needed senators and representatives who were sympathetic to the sugar growers. "Fortunately, sympathy was still easy to buy in Washington; all it took was campaign contributions," suggests the author.

Back to real life. "I'm a fiscal conservative," explained Miller, rocking forward in his chair, his fingers arched together like a priest in earnest conversation. "I look at the sugar program as bad economics. It is big government at its worst." In 1996 Miller cosponsored an amendment to the Freedom to Farm Bill, one of the main objectives of which was to wean farmers off subsidies. The amendment would have begun the process of phasing out the price-support program for sugar, and it had the backing of a broad spectrum of interests. Environmentalists wanted to see the end of the program. A decline in the price of sugar, they reasoned, would mean less land under sugar, and therefore less environmental damage to Florida's Everglades, one of the world's greatest wetlands. The major industrial users of sugar and sweeteners—Coca Cola, M&Ms, Bobs Candies, and the like—wanted the program to end so that they could buy cheaper sugar and sweeteners. And proponents of a free market backed the amendment for ideological reasons. The measure was defeated by 217 to 208, but it would have succeeded had four Democrats and two Republicans, all cosigners of the amendment, not switched sides at the last moment. Four members of Congress—all Democrats—also failed to vote.

Miller was too much of a gentleman to say so, but the Democrats may well have come under pressure from the White House, whose failure to back the amendment may have had something to do with a phone call that President Clinton received shortly before the vote —a call that only came to light during Kenneth Starr's investigation of the Lewinsky affair. Apparently, the president had just told Monica Lewinsky that he was unhappy about their relationship—he was prepared to hug, but not kiss her—when he received a call from Alfonso Fanjul, joint-owner of Flo-Sun, parent company of Florida Crystals, one of the biggest sugar companies in the United States. They spoke for 20 minutes. Perhaps the president and Fanjul, one of four Cuban brothers who fled Castro's regime in the 1960s, discussed the weather. More likely, Fanjul, whose company had made large contributions to the Democratic party, impressed upon the president the importance of the sugar program. He may also have objected to Vice-President Al Gore's announcement that he wanted Florida's sugar growers to pay a 2 US cents a pound tax to help clean up sugar-related pollution in the Everglades.

But why, I asked Miller, were so many politicians in favor of a program that only benefited sugar farmers, of which there were not many? "Because the agrics stick together," replied Miller. "When I went to the agricultural committee to explain my scheme, you could identify everyone and what crop they were representing." He motioned as if to an assembled committee: "There were two sugar guys; and there were two peanut guys. And they were on the committee to defend their particular product." The peanut pro-moters, explained Miller, would always support the sugar program and, in return, the sugar promoters would always support the peanut program. And the same went for the other program crops—cotton and tobacco—that were also beneficiaries of price-support and loan programs.

"And don't forget the corn guys," added Miller with a wry smile. Around half of the natural sweetener produced in the United States comes from sugar beet, a root crop that grows in temperate climates, and sugar cane, a bulky grass ideally suited to the sub-tropical conditions that prevail in the Southern states. The rest comes from corn—750 million bushels of it, grown in 20 states and processed in 17 refineries. The price of corn sweetener is tied to the price of sugar. If the latter fell, then the price of corn sweetener would also fall. This would not be a happy situation for farmers in the Midwest. So the sugar program, which keeps prices high, is in their best interests, too.

Soon after I left Miller, I flew south to the sunshine state, and I began my travels in Orlando, where I met Charles Lee, the vice-president of the Florida Audubon Society. A short, grey-haired man whose soft way of speaking seemed at odds with his uncompromising opinions, Lee didn't have a good word to say about the sugar industry. In his view, sugar cane farming was the primary source of the pollution that threatened the survival of the Everglades. Besides releasing phosphorus into this watery paradise, and in places turning a biologically rich wetland into a cat-tail monoculture, "Big Sugar" had had a dramatic influence on the region's water supply. During the winter dry season, when water was a precious commodity, sugar farmers sucked it out of the system to irrigate their fields. During the wet season, water was expelled from the fields back into the wetlands. So, more often than not, the Everglades either had too much water, or too little.

But it wasn't just the Everglades that were suffering, said Lee. Taxpayers and consumers were also losing out. Local taxes financed the organization who pumped water in and around the system, and local taxes were being used to clean up the mess largely caused, according to Lee, by sugar cultivation. Vast quantities of both federal and state money—from US$8–US$11 billion—were also about to be deployed in what was be the largest restoration project of its kind anywhere in the world. Inextricably linked to all of this, to both the damage and the repair work, was the sugar program, which was also keeping the price of homegrown sugar at double its world market value. Great for sugar farmers; a pain in the wallet for consumers. Lee cited one government report that suggested that the sugar program led to US consumers paying US$1.4 billion more each year for their sugar-infused goods than they would if sugar was purchased freely on the world market. Since then another report has put the figure at US$1.9 billion. He had no doubt that it was the sugar program that had made the crop so profitable and had led to a rapid increase in the area under cultivation in Florida, from less than 180,000 acres when the present program began in 1981 to a little short of 500,000 acres today.

With Lee's stinging critique of the sugar industry ringing in my ears, I headed south, towards the great bowl of sugar land north of the Everglades. The journey was remarkable for nothing other than its monotony. The flat cattle pastures on either side of Florida's Turnpike were interspersed with the odd patch of scrub forest and numerous billboards promoting cheap deals in Walt Disney World and Miami, and the vistas scarcely improved along the narrower roads that took me south-west towards Lake Okeechobee. Years of cultivation had caused the soils around Okeechobee to sink—they are still sinking at the rate of 1 inch a year—and the lake was mostly hidden behind high levees.

"If you've got time," Lee had suggested over a Chinese lunch in Orlando, "you should take a look at Belle Glade. There are parts of that place that are so poor you'd think you were in Port-au-Prince, Haiti. The sugar industry is always saying, 'If you enviros get your way, it'll cost those poor people their jobs.' But they get rid of their workers at the drop of a hat."

I didn't stop this time—iodine-black clouds, forked with lightning, were rolling in across the sugar cane fields, driven by a gusting wind—but the following day I returned with Judy Sanchez, US Sugar Corporation's feisty communications director. A native of Belle Glade, she was incensed by Lee's slur on her town, and she drove me down Fifth Street, which is about as bad as Belle Glade gets. "I'm not saying we want America to be like this," she said. "But it's certainly not Port-au-Prince." A local pizza company—now out of business—would never deliver around this shabby, litter-strewn, much boarded-up neighborhood, and you certainly wouldn't want to wander down these streets after dark.

But was it not true, I asked, that several years ago US Sugar had thrown 1500 people out of work overnight when it closed down South Bay Growers, the biggest vegetable producers in this area? "I was one of those workers," said Sanchez, challengingly. Since the North American Free Trade Agreement (NAFTA) had come into force, cheap Mexican vegetables had swamped the US market. That year prices were so low that South Bay Growers had to plow over one third of their crop back into the fields. Closure was an economic inevitability. "The company did everything it could to help its workers find new work," said Sanchez.

So what about the guys who were hanging around in Belle Glade's back streets? Sanchez gave an oblique answer: "There's jobs at US Sugar for those who want them." Later she took me to see the manager of US Sugar's new refinery. He looked and spoke like the Hollywood image of a good old boy: he had a lived-in face on a short neck, a manner which oscillated between the irascible and charming, and the confidence which comes from getting one's own way. "There's plenty of work here for those who want it," he growled, "and if they don't have the skills we'll train 'em up." Earlier in the month they had advertised 50 new jobs. They'd only had a handful of takers. He said the wages were good, too, but many preferred to live on welfare rather than do a proper job.

Beyond Belle Glade the road curled round the southern end of Lake Okeechobee through much diked sugar cane fields to Clewiston—"The sweetest little town in America," according to the sign on the way in. Clewiston virtually owes its existence to US Sugar, which set up here in 1927, and at its heart sits the handsome Clewiston Inn, owned by US Sugar like virtually everything else of any consequence. Just in case you are in any doubt about the importance of the crop, you are handed a bag of cookies as you head up to your room. The label reads: "Please enjoy our complimentary sugar cookies, baked with pure sugar cane sugar produced by the 2800 employee-owners at US Sugar Corporation. A sweet thank you for choosing us and sweet dreams."

For an hour or so the following morning Bob Buker, the vice-president of US Sugar—the main competitor of the Fanjul brothers —talked about the economics of sugar growing, about the need for a sugar program, about the disgraceful dumping of subsidized sugar by Europeans, and about the sugar industry's efforts to cooperate with the state authorities in cleaning up its pollution. Buker was a small, rotund man and he possessed the fastidiousness that is common among small men: every now and then he would take a comb from his pocket and run it through his hair. At times he sounded weary, almost bored. These were views that he had obviously expressed many times before. A former helicopter pilot in the US army, Buker had a reputation as an aggressive, in-your-face defender of the sugar program, but there was little in his demeanor to suggest a quick and fiery temper. Until, that is, I mentioned Charles Lee's claim that the sugar program had encouraged the cocaine trade.

"What?" screeched Buker.

I recounted Lee's contention that the sugar program had done much harm to Caribbean and South American farmers; a key plank of the program was a strict quota on imports. "One of the effects," Lee had said, "is that folks in South America have been left with just one crop to grow. It's called cocaine, and they ship it to the United States, and our young people die as a result."

At the time this had struck me as a preposterous claim. Buker agreed. "Oh, that is so stupid!" he shouted, banging the boardroom table with his fist. "I'll say it again, categorically, that Charles Lee is a liar. The reason Colombia sells cocaine to the United States is because cocaine is extremely valuable. Sugar sells here—and the Colombians have a quota—at 22 US cents a pound. Last time I checked, cocaine was worth considerably more than 22 US cents a pound." He added that if the United States unilaterally did away with the sugar program, then Colombia would have to sell its sugar to the United States at the world market price of around 10 US cents a pound, less than half of what it currently received under quota.

"Charles Lee has even blamed us for AIDS in South Florida," said Buker with an ironic laugh.

"Why? What have you been doing?" I asked.

"I don't know. He will say and do anything to hurt farmers."

The sugar growers are out to grow as much sugar as possible for as great a profit as they can get; the environmentalists want to see as little sugar grown as possible at zero impact on the environment. Both sides have well-oiled propaganda machines, and are adept at emphasizing the facts that suit their argument, while burying from sight those that don't. And both make much of their own honesty, while deploring the mendacity of their opponents. At one point Buker said: "Charles Lee is telling you a true fact in order to deceive you." And Lee said, apropos a claim by the sugar industry: "Figures don't lie, but liars figure." When I repeated these counter allegations to an ecologist in Miami, he smiled and responded with an aphorism: "All stories are true. Some of them actually happened."

The first thing we need to establish is what exactly happened to the Everglades. The swamps, as I was to discover when I visited the territory of the Miccosukee Indians, have been degraded; but how much is down to sugar cultivation, and how much to other factors such as urban development? This is the key environmental issue, but the economic issues are just as important. Is "Big Sugar" paying its fair share of the clean-up operation that is now in progress, or is the taxpayer having to subsidize the clean-up of what, in effect, is the mess created by sugar cultivation? Finally, is the sugar program really making sugar twice as expensive as it would be if there were a free market in sugar in an unsubsidized world, as Lee and others claimed?

* * *

At first glance there is nothing to suggest that the lives of the Miccosukee Indians who live along the Tamiami Trail differ much from the lives of thousands of other Americans who inhabit small strung-out highway settlements. Their concrete block houses are surrounded by the familiar detritus of modern life—satellite dishes, barbecues, pick-up trucks, trailers, children's bikes, itching dogs. But pull off the main highway and you suddenly come across a cluster of smart modern buildings, all constructed with the profits made from the tribe's gambling operations. There is a new secondary school, a gymnasium fit for a professional ball team, a court house, a center for the treatment of drug addicts, a police station with enough cars to patrol a small town, and an administration block—all of this for a tribe who numbers less than 600 souls.

The administration block was opulent and airy, and in the foyer two strikingly beautiful Indian women, each armed with a young baby, were discussing tribal affairs. A surly receptionist, who happened to be white, told me to sit down and I sunk into a comfortable chair with a magazine and waited for Gene Duncan, who heads up the Miccosukees' water resources department. I had pictured him as someone who would look like a scientist, with a studious air about him, dressed perhaps in a white laboratory coat, and was taken aback to be greeted by a large character with a tremendous belly. He was kitted out like a Hell's Angel with all the emblems of bikerdom: skull bracelets, black tee-shirt, and some serious tattoos, one of which spelt the name the Indians had given him. Translated, it read "man with red hair," which was scarcely true now, as he had lost most of it.

"The Miccosukee have a very non-confrontational nature," explained Duncan as we headed for the water's edge. They employed people such as him to deal with the outside world, a world that had frequently betrayed them in the past. During the early 1800s, the US army drove the Indians into the swamps; over a century later they were pushed onto the Tamiami Trail when the federal authorities established the Everglades National Park. When they arrived, the Everglades stretched from the tip of Florida to present-day Orlando. Today they cover less than half of this area and Lake Okeechobee, which used to occupy the heart of the Everglades, now lies to the north and is separated from the portion that the Indians occupy by sugar cane fields.

"So you're their confrontation?" I suggested tentatively to Duncan as we climbed onto an air boat.

He laughed. "Yes, sir. That's right." Since Duncan had arrived, the Miccosukee had sued the state and federal governments—generally on the issue of water quality—some 14 or 15 times. "When we go into the courtroom now the whole front bench is lined with attorneys, and some of these will be ours," he explained with relish. "We're up with the big boys now!"

We headed out into the swamps and came to a halt near a thatched *chikee,* one of the many abandoned dwellings where the Miccosukee used to live. Such was the noise made by the air boat, a contrivance that consisted of a shallow boat propelled by an engine big enough to power a small airplane, that the silence of the swamps, once the engine was turned off, was almost shocking. Neither Duncan, nor I, nor the boatman—a white American who recently married into the tribe—talked for a while. Instead, we drank in the primeval wildness of the landscape. As far as the eye could see the land was either leased in perpetuity or owned by the Miccosukee. Altogether, they presided over some 300,000 acres.

It was time, now, to get down to some serious chemistry. Duncan estimated that the phosphorous levels in this pristine patch of the Everglades were around 6–10 parts per billion (ppbn), and in this nutrient-poor environment nature thrived as a mosaic of short sawgrass, scattered lilies, and bladderwort, and shallow slews of open water. Just below the surface we could see the thread-like web of spongy periphyton. "The periphyton is the key," said Duncan. A mix of algae and bacteria, this is the basis for the whole food chain. The microscopic organisms that feed on the periphyton are food for fish and amphibians, and these in turn sustain the rich bird life, not to mention alligators, snakes, and other creatures.

Further north, towards the sugar cane fields and the Everglades Agricultural Area (EAA), the picture is very different, largely as a result of the phosphorus that is leaching out of the agricultural area. "In those conditions," explained Duncan, "there is no periphyton and the sawgrass grows to 12-foot high." At its worst, where phosphorus levels are highest, the wetlands have been taken over by cat-tails. Whenever this has happened, water birds and most other creatures have disappeared.

So was the sugar industry to blame? "I don't know whether I could blame sugar," drawled Duncan in his lyrical Kentucky accent. "They were invited to come, to help make the area prosper. Of course, it's had terrible consequences. But it's not just about sugar. It's about having a piece of property that's supposed to be under water being exposed to air. Matter of fact, sugar is quite a clean crop. It would be much worse if it was all down to vegetables." All the same, the water coming off the cane fields and onto the Miccosukee land contains around 100 ppbn of phosphorus. "And that's ten times more than it should be," said Duncan.

There is a certain inevitability about all of this. The drainage and cultivation of these soils causes them to oxidize, and the main source of phosphorus in the surrounding wetlands is not artificial fertilizer, but this slow decay of the soil. There is considerable debate, however, about how much of the phosphorus comes from agricultural activity and how much from other sources. According to Charles Lee, 80 percent to 90 percent of the phosphorus in the system can be traced back to the EAA, and virtually all of this is due to sugar cultivation.

Not true, said Bob Buker. It may well be that most of the phosphorus in the wetlands to the south of the EAA comes through the EAA, but sugar growing is only responsible for around half of this. The rest comes from urban and agricultural development to the north of Lake Okeechobee, from as far afield as Orlando, and simply passes through the cane field canals on its way south. Empty your bowels in Disney World and you, too, can contribute to the degradation of the Everglades.

"I cannot agree with that," said Mark Kraus, an ecologist with the National Audubon Society in Miami, "and, in any case, for years the sugar industry back-pumped waste water into Lake Okeechobee, so Okeechobee's pollution is partly their responsibility." There is also disagreement about what the phosphorus levels should be in the Everglades. Duncan, Lee, and Kraus were adamant that the natural level of phosphorus is around 10 ppbn. In contrast, a researcher hired by the sugar industry has claimed that 20–30 ppbn is the norm.

This being the United States, there has been endless litigation; and it was the fuss over phosphorus that sparked off a long series of lawsuits, beginning in October 1988 when the US Department of the Interior brought a suit against the state of Florida. Its contention was that the South Florida Water Management District was failing to reduce the pollution entering federally managed areas. Several years and US$25 million of taxpayers' money later, a settlement of sorts was reached. It was agreed that farmers in the EAA would adopt best management practices to reduce phosphorus run-off by 25 percent; that limits would be set for phosphorus levels in the national park and the wildlife refuge; and that 35,000 acres of "storm-water treatment areas" would be built around the EAA to reduce phosphorus from agricultural run-off. Two years later, in 1993, a statement of principals was negotiated by the sugar industry and the state and federal governments. It was said at the time, by the secretary of state for the interior, Bruce Babbitt, that the sugar industry would pay the "lion's share" of the clean-up cost.

The plan was given legislative force by the 1994 Everglades For Ever Act.

In Lee's view, the storm-water treatment areas are working well. Phosphorus levels have been reduced to 30–50 ppbn from around 100 ppbn. Lee and Duncan both admitted that the sugar industry's best management practices are helping, too, although they were not as bullish in their praise as the sugar growers were of their own achievements. "We were given a mandate by state law to reduce the phosphorus by 25 percent," said Buker, "and we've done twice as well as the law requires." On average the farmers in the Everglades Agricultural Area have reduced the amount of phosphorus leaving their land by 50 percent. My visit to a sugar farm—admittedly, it was chosen by US Sugar—convinced me that the company has achieved much in terms of reducing its phosphorous run-off; but as the environmentalists pointed out, reducing levels from 200 to 100 ppbn sounds more impressive than it is. Assuming that natural levels are as low as 10 ppbn, there is still a long way to go.

But is the sugar industry paying its fair share of the clean-up? Yes, said Buker: "We're paying 100 US cents in the dollar for cleaning up *our* own water." The first stage of the clean-up was costing around US$700 million, and the sugar industry was paying approximately one third of this. None of this impressed Lee, who maintained that the taxpayer was having to pay two-thirds of the clean-up bill, and what was being cleaned up was primarily sugar's mess. It was the belief that "Big Sugar" should pay more that led to the most expensive political campaign in Florida's history. Lee suggested that I should get the full story about this seminal event in the history of the long struggle against "Big Sugar" from one of its key participants, Mary Barley.

It was a relief to escape sugar country. With its vast fields of cane and its dikes laid out with mathematical precision, with nothing to break the monotony of the flat landscape save for a few towering factories and ugly strips of fast food joints and gas stations, this was a world devoid of any aesthetic appeal. The land seemed soulless to me, and no more natural, but considerably less appealing, than Miami's South Beach, where I passed the night on my way south to see Mary Barley in Florida Keys.

When I arrived at Mary Barley's secluded home overlooking Florida Bay, I was greeted by her South American maid and two dogs that looked like the canine equivalents of a couple of teenage girls heading out for a night on the town. They were slight creatures, excessively well groomed, obviously spoilt, and eager for attention. Mary Barley duly appeared and immediately set about making lunch. I imagine she was about fifty; she had dark hair, strong features, and a business-like way of moving around. She was both good company —her tongue was as sharp as her homemade horseradish sauce— and a good cook. Until the wind drove us indoors, we sat by the swimming pool and ate fresh tuna, seared on the outside and raw in the middle, and succulent prawns, washing them down with an excellent white Rioja.

This classy treatment was not what I had come to expect when entertained by environmentalists; but, then, Mary Barley was no ordinary environmentalist. Her swift transformation from socialite wife of the property developer George Barley to anti-sugar campaigning widow has been well documented by the local media. Prior to his death in a plane crash in 1995, George Barley mounted a vigorous crusade against the sugar corporations, whose demand for water, he believed, was leading to the biological impoverishment of Florida Bay. Following his death, his abrasive widow took up his cudgels— she now runs four groups whose mission is to "save" the Everglades —and in early 1996 she and others began to plan a ballot initiative, which, if approved by the voters, would have led to a penny-a-pound tax on sugar farmers. This tax, like Al Gore's, which the Democrats quietly abandoned, was to go toward Everglades restoration.

"At that time," she recalled, "Newt Gingrich held all the power in both the Senate and the House. He liked the sugar issue and wanted to counter Clinton and Gore. He and his Republican friends also wanted us out of their hair and they started negotiating. They said, 'What if we give US$200 million to the Everglades out of the Farm Bill?'" In return, Mary Barley—a Republican—and her environmental colleagues would drop the ballot initiative. "And you know what?" she said, rolling her eyes in mock incredulity. "They were literally going to take the money out of school lunches! We were so shocked. We could not believe they'd say that. So we said no, we'd take our chances." In the end the Republicans diverted US$200 million from farm spending for Everglades restoration, but Mary Barley and the environmentalists still pressed ahead with their ballot initiative.

In Florida the state constitution can be changed in two ways: by a vote within the legislature, or by a vote on a constitutional amendment placed on the ballot at the time of a general election. In 1996 the environmentalists succeeded in getting three amendments on the ballot. Two of these—one establishing the principal that polluters should pay to clean up their pollution; the other calling for the setting up of a trust fund to manage finances for Everglade restoration—were successful. The key amendment, however, to impose a penny-a-pound tax on the sugar industry, which had it been successful would have raised around US$35 million a year for Everglades restoration, failed.

"Penny-a-pound failed because they lied," explained Mary Barley bluntly. "They said it was a tax on people. They even got Jesse Jackson telling black people that their medicine would go up. I'll guarantee that the sugar industry spent US$50 million on that campaign, even though they only reported US$30 million." The pro-tax campaigners —led by the Audubon Society and Mary Barley's organizations— also put their money, or rather the money donated by commodity broker Paul Tudor Jones, where their mouths were. Jones was an old friend and fishing pal of George Barley, and he had a holiday home in the Keys. The environmentalists spent an estimated US$15 million trying to convince Florida's voters to approve the amendment.

The environmentalists claimed that the sugar industry's Citizens to Save Jobs and Stop Unfair Taxes campaign was a masterpiece of misinformation. For a week prior to the vote, all sugar installations and factories were closed down, and the workers, at least half of whom were black or Hispanic, were sent out on the campaign trail. Kraus recalled pro-tax campaigners being shouted at and called racists by sugar workers at a voting station in Miami. "It was simply not true," he said. "Race had absolutely nothing to do with it. This was about a tax on sugar." Besides mounting a vigorous TV campaign that claimed that this was a tax on the people of Florida, not on sugar, the sugar industry targeted large numbers by post, making a similar claim. Their message was: this tax will hit you voters where it hurts, by adding to your taxes.

Bob Buker's justification for this line of argument was intriguing, if not plain disingenuous. Briefly stated, he claimed that a tax on sugar would bankrupt the sugar industry in five years, and that the public would have to pay taxes to clear up the financial mess this left behind. Therefore, a penny a pound tax on sugar was a tax on the public. It seemed that the majority of voters believed this.

According to Buker, the profit on sugar was less than a penny a pound. Fair enough, said the opponents of "Big Sugar": prove it by showing us your books. The sugar industry has consistently refused to do this, and we are therefore left with the economic assessments made by outsiders. I take one, almost at random, from a pile of reports on my desk. Four agricultural economists analyzed data from the US Department of Agriculture's (USDA's) *Sugar and Sweetener* publication. During 1985–1994, returns per pound of sugar varied between 2.8 and 7.5 US cents. Even with the penny a pound tax, the industry could expect returns of around 4 US cents a pound. The claim that the industry could not afford to contribute a penny a pound tax was clearly a lie. It could, but then its profits would be less.

One of the reasons why the sugar industry makes such handsome profits is because the US sugar program ensures that sugar growing is highly profitable. Rigoberta Lopez, an economist at the Department of Agriculture and Resource Economics at the University of Connecticut, estimated that the benefits of public policy to the

Florida sugar producers amounted to US$3.4 billion between 1970 and 1994, or US$137 million a year. The bulk of these benefits came from the sugar program's price-support mechanism, rather than through water subsidies. The significance to the industry of the sugar program can be gauged by the vigorous way in which companies such as the US Sugar Corporation and Flo-Sun campaign for its continuation. So, it is undoubtedly in their interests. But what about the consumer?

A combination of loans, price supports, and import quotas has kept the price of sugar in the United States at around 21–23 US cents a pound, double its world market value. Consequently, says the US General Accounting Office (GAO), consumers are paying an extra US$1.9 billion on their food bills. But they're missing the point, say the sugar growers. The sugar program has helped to keep prices stable and the US consumer gets an excellent deal. While sugar costs just 39 US cents a retail pound in the United States, it costs 68 US cents in France and US$1.04 in Japan. "I could courier sugar to Paris and it would still be cheaper than the sugar the French farmers produce," maintained Buker, and he meant it.

The GAO report assumes that consumers would be able to buy sugar at the world market price of around 10 US cents a pound. This is nonsense, said Buker, because the world market price of sugar is not the price that sugar would be in an unsubsidized world. The reason why it is 10 US cents a pound—and at that price no one can make a living out of sugar, according to Buker—is because the European Union (EU) dumps around 6 million tons of sugar on the world market each year, or approximately two-thirds of the quantity consumed by the United States annually. The producer subsidy provided to EU farmers has led to a massive increase of the sugar beet crop. This, in turn, has created surpluses that must be stored at the taxpayers' expense, and the dumping of these surpluses on the world markets has inevitably depressed prices. "The problem is Europe," said Buker emphatically. "Europeans are rabid about keeping their agricultural programs."

Contrary to claims made by the sugar industry, there is no doubt that sugar farmers in the United States are pampered by the government, even though the 1996 Farm Bill abolished certain privileges, such as guaranteed minimum prices, and increased the quota for imports. The bill stipulated that sugar farmers could apply for non-recourse loans—which means that farmers can repay the loans in sugar, rather than in cash—providing imports remained over 1.5 million tons a year. In 1999 they fell to 1.25 million tons, yet the USDA continued to provide non-recourse loans. What is more, in 2000 the US government paid sugar producers over US$400 million of taxpayers' money for forfeited and surplus sugar. However you choose to perform the calculation, the main beneficiaries are corporate agriculture and the big sugar farmers; 42 percent of sugar program benefits go to just 1 percent of sugar growers, with the 33 largest annually receiving over US$1 million each in benefits. One family, the Fanjuls, gets US$60 million more a year than it would if the program did not exist.

Charles Lee was somewhat overstating the case when he claimed that parts of Belle Glade were like Port-au-Prince, but you don't need to drive far off the main highways in this part of Florida to see serious poverty. I recall, in particular, driving through a small town at around 11 o'clock one morning on the east of Lake Okeechobee. The population in one area was almost exclusively black. The housing was decrepit and shabby, and judging by the large number of adults on the street a good portion of the population was unemployed. There was a look of poverty about both the place and the people; although I was glad to see that some of the men were playing cricket, an English game imported to Florida by their Caribbean ancestors, who came here to work in the sugarcane fields. These people, like everyone else in America, were paying far more for their sugar than they would if there were no subsidies, and their annual food bills were at least US$350 more per person than they would be in an unsubsidized world. This might be small change to many Americans, but it isn't for people like these.

True, Bob Buker was probably right when he said that if subsidies to sugar farmers ceased worldwide, then dumping would cease, and the price of sugar would rise well above 10 US cents a pound, so consumers would not save US$1.9 billion a year. Still, the fact that the corn producers and sweetener manufacturers are so keen to keep the sugar program—their prices are pegged to those of sugar—suggests that the price would be lower (probably considerably lower) than it is in the United States at present. Whether the industrial users of sugar would pass on the savings to consumers remains to be seen, but the price of a bag of sugar in your local store would undoubtedly be less.

If Europe and the United States abandoned their policy of supporting sugar farmers, there would be many more winners than losers. Consumers on both sides of the Atlantic would be better off, as would taxpayers. Producers in poor countries would be able to compete more effectively with agribusinesses such as US Sugar and Flo-Sun, and a downturn in prices in the United States and Europe would lead to land coming out of sugar production. In Florida, environmentalists speculated that the area under sugar would shrink, and that this would be to the benefit of the Everglades. They may be right, although farmers might decide to replace sugar with crops that would lead to just as much, or more, phosphorus run-off. Whatever the outcome, there would still be no sensible justification for retaining sugar subsidies. Bob Buker said that it would be unfair if the United States acted unilaterally and abolished its sugar program while subsidies remained in place in Europe. In a sane world, both the United States and the European Union would abolish these and all other production subsidies, and do away with quotas and tariffs such as those which protect the US sugar industry.

I note, in passing, that President George W. Bush has made known his position on quotas. During the 2000 presidential election campaign, he was quoted in the *Economist* as saying: "What I am against is quotas. I am against hard quotas, quotas they basically delineate based upon whatever. However they delineate, they vulcanize society. So I don't know how that fits in with what everyone else is saying, their relative position, but that's my position." Quite what this means I am not sure: your guess is probably as good as his.

*   *   *

The US sugar program is not the only crop program that raises prices for consumers, discriminates against foreign growers and grabs a significant slice of the taxpayer cake. Nor is it the most egregious, according to a White House economist whom I met on my travels. He was incandescently angry about the peanut program, because it penalized those who could least afford high food prices. During the time of writing, the peanut program paid farmers US$610 a ton, almost twice the world market price, for a government-set poundage quota. This meant that US consumers paid around US$500 million a year in higher prices. This was especially hard on the poor, for whom peanut butter is an important staple. The 2002 Farm Bill introduced a new plan for peanuts. This involves the government buying up peanut quotas. Those who own them can continue to grow peanuts under a new subsidy system. According to the USDA, a farmer with 50 acres of peanuts, producing a ton an acre, will now get US$694 a ton. The quota buy out will cost taxpayers an estimated US$1 billion.

The cotton program, like the peanut program, was set up during the 1930s to help alleviate poverty among farmers during the Great Depression. Times have changed, but the program still raises the price of cotton way above its world market value and helps the cotton industry compete—unfairly—on the world market. The 2002 Farm Bill has effectively doubled cotton subsidies. Analysts suggest that this will almost undoubtedly lead to an increase in US cotton exports, and this could have a catastrophic effect on cotton producers in West Africa and other parts of the world. Cotton, like peanuts, requires heavy irrigation and liberal doses of agro-chemicals, which can be damaging for the environment.

The tobacco program was also established during the 1930s; its aim was to stabilize prices by controlling the amount of tobacco coming onto the market. According to the *2001 Green Scissors* report, the government spends at least US$100 million a year assisting tobacco producers. The environmentalists point out that the taxpayers should not be asked to fund the growing of a crop that makes intensive use of pesticides, fertilizers, and energy. They also spout much sanctimonious clap-trap about smoking being bad for our health. This is none of their—or anybody else's—business, and, in any case, revenues from tobacco taxes exceed public expenditure on tobacco-related diseases by a considerable amount. Smoking is good for the Treasury; the tobacco program, in contrast, is not.

The environmentalists make much of the fact that the cultivation of these program crops involves the use of massive quantities of pesticides, herbicides, and artificial fertilizers. Taxpayers and consumers are therefore being involuntarily inveigled into paying for activities that damage the environment. This is quite true; but then much the same could be said for the subsidies that go to farmers growing corn, wheat, rice, and other crops, and about which the environmental campaigners have relatively little to say.

A significant portion of agricultural support comes under the heading of "General Services", and includes government spending on research and development, inspection services, and marketing and promotion. One of the subsidies that falls into this category is the Market Access Program (MAP), which has cost US taxpayers US$1.5 billion over the past decade. MAP's purpose is to encourage exports of agricultural produce, and it does this by paying for the advertising campaigns, trade shows, and consumer surveys that help to boost sales abroad. The main beneficiaries of MAP have been corporations such as Sunkist, Ocean Spray, and the American Forest and Paper Association. For the year 2000, this subsidy was set at US$90 million. By 2006 it will amount to US$200 million a year.

The problem with many subsidies is that they encourage overproduction. If anyone doubts this, they need only look at the sad saga of oilseed subsidies in the United States. In 1997 government support accounted for some 4 percent of oilseed producers' incomes. This figure rose to 15 percent in 1998 and 23 percent in 1999 and 2000. Total government support to oilseed producers was set at US$4 billion for 2000, or around US$11,250 for every oilseed farmer. These massive subsidies help to explain why US oilseed producers increased the area they planted by an average 6 percent a year at a time of low world market prices. Such fiscal madness is commonplace, and a ten-year-old with only the most rudimentary understanding of economic theory could justifiably think that our legislators have taken leave of their senses.

The 1996 Farm Bill made a laudable attempt to reduce farmers' dependency upon the taxpayer. The bill struck out a portion of the federal program whose purpose had been to manipulate production and prices. In its place, farmers were given direct payments that were designed to decrease over a period of time. During the mid-1990s, all seemed rosy for US farmers; but by the end of the century, with the dramatic decline in commodity prices and loss of export markets, life didn't seem so good, and farmers returned once again to Washington, DC, with cap in hand. In 1999 farmers were given a US$6 billion emergency bail-out, and in June 2000 the Agricultural Risk Protection Act set aside a further US$15 billion in taxpayer-funded farm relief.

The 2002 Farm Bill sanctioned an extra US$80 million of subsidies over the next ten years. This represents an increase in government farm spending of around 80 percent. The Bush administration has argued that as bail-out packages averaged some US$7.5 million over the previous four years, the bill simply continues current spending practices. But this misses the point. The fact is that the Farm Bill represents a massive repudiation of the prudence that its predecessor was designed to encourage. The major bene-ficiaries of the new bill will be arable farmers in the prairie belt, with half of all new subsidies going to just six states. The biggest winner of all is Texas. Direct crop subsidies to the president's home state will double during the first year of the new bill. The losers, needless to say, are US taxpayers and foreign farmers, who must now compete on world markets with heavily subsidized US exports.

"Today's subsidies have become rewards to the economic victors as a form of insult to the defeated," suggests one American commentator, and this is as true for Europe as it is for the United States. Europe, as Buker said, is rabid about its subsidies, around two-thirds of which go towards supporting producers of sugar, cereals, oil seeds, tobacco, and other crops. It is the big farmers who have benefited most, as payments are almost always tied to the areas sown or quantity produced. In the UK, between 1993 and 1998, 327 farmers received the equivalent of over US$1.6 million in subsidies, 27 received over US$4 million and 6 over US$8 million. Goliath has won, David has lost, and the environment has suffered, too.

*   *   *

This chapter began with politics, and it is to politics that we must return if we are to understand why politicians continue to vote for programs that are in the worst interests of consumers, taxpayers, and the environment. Many politicians would answer by saying, among other things, that these programs help to provide jobs and stimulate economic growth. The specter of unemployment haunts politicians, and in democratic societies the loss of jobs within a constituency has often proved to be the last nail in the coffin of struggling politicians. For purely selfish reasons, then, politicians may view measures that create or maintain employment as worthy of their support, and this goes some way towards explaining why the recipients of subsidies play the jobs card for all it is worth.

The US sugar policy, says one of the industry's glossy brochures, "creates 420,000 jobs and adds US$26.2 billion annually to the economy." The implication is that a significant number of jobs would be lost if the program ceased. However, the claim that 420,000 jobs could be at risk is ludicrous. A study relying on data from the

USDA suggests that the number employed full time by the sugar industry is closer to 46,000. Furthermore, since the present program came into force in 1981, over half of the sugar refineries in the United States have been forced to close down. Why? Because they could not afford to buy sugar at the prices inflated by the sugar program. Over 1000 jobs alone were lost to two Philadelphia mills. Guy McCormack, the president of Bobs Candies, the largest US candy-cane manufacturer, was quoted in an article given to me in Florida as saying: "Our company would save US$2 million a year in raw-materials costs if the sugar program were eliminated. That would help us keep jobs in America, and it would lower the retail price of our candy-canes by 10–15 cents a package." Bobs Candies recently opened a new plant in Jamaica, where it can buy sugar for half the US price. So the jobs argument, as far as the sugar industry is concerned, is weak at best.

The jobs that really matter, as far as many politicians are concerned, are not so much their constituents', as their own. They have a vested interest in keeping "Big Sugar", and many other business lobbies, happy. Why? Because they need their money. In the United States a politician without campaign funds and a party without a fat bank balance are doomed, and while I was in Florida I was given an insight into this seedy world by an attorney who had observed how the sugar industry used its wealth to influence the political process. He was a Republican with a keen interest in public education and environmental issues, concerns traditionally considered the provenance of Democrats. As soon as we sat down in a restaurant frequented by smart-suited business types, the attorney—he asked me not to mention his name—explained how corporations, and others for that matter, contributed to political campaigns.

First, there was "hard" money. Over an election cycle, individuals could give US$2000 to a candidate they favored: US$1000 for a primary and US$1000 for the general election. Political action committees (PACs) were limited to donations of US$10,000 per election, per candidate: US$5000 for the primary and US$5000 for the election. This may sound like peanuts. It wasn't, according to the attorney. PAC donations alone amounted to around a quarter of the US$791 million raised for House and Senate campaigns during the 1995–1996 election cycle.

"Then you've got independent expenditure and soft money," continued my companion. "I'll give you an example of independent expenditure," he said. "Let's say there's a candidate that sugar don't like—he's against the program. They can put as much money as they want into attacking him in television adverts, and it's all legal so long as they don't coordinate with that person's political opponents." At that time—this was well before legislation to reform the campaign finance system was passed in March 2002—there were no caps on either independent expenditure or on soft money, the donations made by corporations, unions, and individuals to political parties rather than specific candidates. During the 1995–1996 election cycle, US$264 million of soft money was donated to political parties. The US Sugar Corporation—a staunchly Republican outfit, according to the attorney—gave solely to the Republican party. Flo-Sun, the Fanjuls' company, favored both parties. During the last election but one, Alfonso Fanjul served on Republican Bob Dole's finance committee, while brother Pepe raised money for Clinton's campaign. Their promiscuity evidently paid off.

State politicians, suggested the attorney, could be much more easily bought. "Big Sugar virtually owns Florida's state legislature," he said. "It's a less expensive prostitute than federal government. Those politicians in Tallahassee make US$22,000 a year and sugar gives them a shit load of money to get them on their side." In Montana I had heard a similar story: ranching money kept politicians sweet. And later, I was to hear the same said about mining interests in Arizona and New Mexico.

During the 1995–1996 election cycle, the Fanjuls, their companies, and their employees gave US$709,000 to federal election campaigns. Overall, sugar producers provided US$2.7 million of election finance. Taken from January 1993 through December 1998, the sugar industry PACs put over US$5.7 million into congressional campaigns, on top of US$1.8 million of soft money. None of the top 40 House of Representative recipients of sugar PAC money voted to end or reform the sugar program. On average, those who voted to retain the program received US$13,261 each, compared to US$1964 for those who voted to end or reform it. Five of the six cosigners of the Miller amendment who switched sides, and voted to retain the sugar program, subsequently received campaign donations from the sugar industry. The member who did not was not standing for re-election.

Since I traveled through Florida's sugar country, matters got both worse and better as far as the campaign finance issue is concerned. In the view of the *Economist*, the 1996 election "supposedly set an all-time low in money-grubbing and rule bending." But the 2000 elections surpassed even those sordid times. Just under US$3 billion was spent on presidential and congressional campaigns—large chunks of which were donated by unaccountable groups—and a further US$1 billion was spent on state contests. One candidate spent US$16 million to become a New Jersey senator, and George W. Bush raised US$100 million for his primary campaign. Soft money donations to the two main political parties doubled to almost US$500 million. "The influence of money could be seen everywhere in politics," railed the *Economist*, "from the lavish feasts that special interests put on for their hired shills during the political conventions, to the free cigarettes that Democratic enthusiasts gave to homeless people in Wisconsin in an attempt to persuade them to vote."

That's the bad news. Here is a modicum of good. In March 2002 a campaign-finance reform bill sponsored by Senators John McCain and Russ Feingold was finally passed. President George W. Bush subsequently put his signature to the bill. As we go to press, it is clear that the bill will be challenged in the courts, but assuming it comes into force, there will be some significant changes. The bill bans soft money—in other words, unregulated donations to national political parties. It also restricts the use of "issue ads" within 60 days of a general election and 30 days of a primary. Big deal? Well, not really. As a sop to opponents of reform, the bill raises the limit on hard money, on individual contributions for a primary or general election, from US$1000 to US$2000. This doesn't sound much until you consider the facts. For the 2000 elections, federal candidates and national parties collected US£2.9 billion. Three-quarters of this—US$2.2 billion—was hard money. The US$100 million that President Bush raised for his primary was hard money. The reforms may be a step in the right direction; but they do not mean that corporations, special interest groups, unions, or individuals will be powerless to influence politicians of political parties.

I should add that the popular media portrayal of the Fanjuls as immigrants on the make with fast cars, fabulous mansions, and the paraphernalia of the super-rich, partly acquired through the clever manipulation of politicians, rather misses the point. The Fanjuls, and the other sugar growers, are not doing anything illegal: they are simply running successful businesses and doing whatever it takes to make them as profitable as possible. "The Fanjuls are really very polished, very sophisticated," said the attorney indignantly when I told him that one of the environmentalists I had met described them as slimy low life. "What sucks," continued the attorney, "is not people like them—the Fanjuls are a much classier act than the US sugar guys—it's the politicians with the backbone of chocolate eclairs who are prepared to take their money and do them favors in return."

I was reminded of something that Senator William Proxmire, the initiator of the Golden Fleece awards for outstanding examples of pork-barrel government spending, said about the relationship between wealth and power. "Money will go where the political power is. Anyone who thinks government funds will be allocated to firms according to merit has not lived or served in Washington very long." In this world, the meek will definitely not inherit the

Earth; rather, they are being forced to pay a tithe to the rich, the powerful, and the unscrupulous, either directly through their taxes, or indirectly by paying more for the goods that they buy.

CHAPTER 4

WASTING WATER

It took me around three hours to drive from San Francisco to Auburn, in the foothills of the Sierras. It was the weekend and the car-jammed highways shimmered in the heat as tens of thousands of Californians headed for the cooler climate of the mountains. By luck rather than design, I arrived in Auburn, a small town whose historic buildings looked more like fakes than real relics of gold-rush days, on the day of the American River Festival.

The festival was held at the Auburn Dam Overlook—a misnomer, as there was still no dam—a couple of miles out of town. By the time I arrived the car park was overflowing, and the air was full of the shouts and squeals of excitable children, and the sound of a jazz band playing. The smells were good, too: the sappy aroma of pine mingled with the sweet, smoky tang of barbecued spare rib. I headed past the food stalls in search of Gary Estes, a prominent campaigner against the Auburn Dam, and found him at a stall that extolled the virtues of the American River, and highlighted the perfidy of those who wanted to dam it. A neatly built man with spectacles, a capacious hat, and a nice line in banter, he showed me plans for the dam and an aerial photograph depicting where it would plug the canyon below us.

"The dam is a project in search of a purpose," Estes explained. Prior to an earthquake that halted construction in 1975, the dam was seen as a multipurpose affair. It would provide water for agriculture, hydroelectric power, a lake for recreation, and jobs for locals. Now, if it were built, its primary purpose would be to prevent the flooding of California's state capital, Sacramento.

At the edge of the canyon there was a small table, pinned to which was a sheet of paper with the title "Auburn Dam: Dimensions of a Mammoth Boondoggle." If built, said the fact sheet, the dam would be the largest in the United States. It would cost US$1 billion. It would be wider than three US Capitols. It would require three times more cement than the enormous Hoover Dam, but would provide only 10 percent of the power and less than 2 percent of the water. This, I discovered later, was the Auburn Dam as it was planned by the US Army Corps of Engineers in 1996, not as it was proposed now, in 1999. No matter: the environmentalists still maintained that the latest incarnation suffered from the same defects. The dam would be massively expensive, and there were much cheaper flood control alternatives that would not lead to the flooding of the American River.

In 1998 the Auburn Dam was among the environmentally damaging, subsidy-driven projects listed in the *Green Scissors* report. It was very much a going concern, and the environmentalists were urging politicians not to fund it. A year later, in 1999, the *Green Scissors* report listed its campaign against the Auburn Dam as a victory, confidently referring to it in the past tense. But still it would not die: midway through the summer of 1999 it was again the subject of consideration in both the House of Representatives and the Senate. Congressman John Doolittle, a powerful supporter of the dam and arch enemy of the environmentalists—or communists, as he likes to call them—had made it clear that the Auburn Dam was his Holy Grail. He would not rest until it was built.

There was a poignant moment shortly after midday when a dozen white doves were released in memory of an anti-dam campaigner who had been struck by a train while cross-country skiing with his wife earlier in the year. There were some good speeches—they were moving without being sentimental—and once they were over I went in search of one of the dead man's fellow activists. With his droopy mustache, shaggy tangle of gray-black hair, and shambling gait, Eric Peach looked like Elliott Gould. He was busy firing some pottery fish in a portable kiln. "Symbolically," he smiled, "we're bringing salmon back up the river." He explained that the Folsom Dam, to the south of Auburn, had destroyed the migratory salmon runs in the mid-1950s. Peach said that the loss of Frank—the anti-dam campaigner—had been keenly felt by all of those here, even though they had not been especially close friends. "We're all glued together by the river," he mused as he rose from the kiln and rubbed sweat off his brow. "We love and respect one another, but ultimately our passion and connection is the American River."

Unlike many rivers in the United States, the American River still runs pretty much as nature intended, at least in its upper reaches. The violence committed against rivers has been astonishing; in parts of the United States the demand for water is now so great that many rivers are no more than dried-up slashes in the landscape, with the merest trickle of water making it to the sea. During the past century, over 100,000 dams have been built in the US, and when archeologists of the distant future come to dig among the detritus of the present era, they will be as struck by the dams, levee-lined canals, and straightened rivers as the paleontologists of today are by the bones of dinosaurs. Scrabble deeper among the archives, and they will be shocked to find that the vast majority of these structures were paid for not by the farmers, industries, and individuals who used the water, but by taxpayers. And while the subsidies tended to benefit the few, often at the expense of the many, they also contributed to a whole gamut of environmental problems.

Peach suggested that if I wanted to get a flavor of the river, and appreciate why it should never be dammed at Auburn, then I should join the rafting trip which local enthusiasts were organizing that afternoon. So I ate some food, drank a bit, chatted to Estes, then signed up for the trip with around 30 others. A battered old bus took us in shifts down a winding road towards the river, some 10 miles further upstream. We had to walk the last stretch through the pine and dry scrub, with insects buzzing around us and a hot sun beating down on our backs. Once down by the water's edge, I sat with Frank's widow, Michael. She was an extraordinary woman. She was devastated by her loss, but spiritually serene, and she told me about Frank, about his death, and about what the river had meant to him. This was a place of singular beauty and it seemed a world away from the monotonous, heavily cropped landscape of California's Central Valley, which lay between here and the Pacific.

Eventually the rafts arrived, and by the time we pushed off the sun had fallen behind the rim of the gorge. The river's passage was frequently interrupted by truck-sized blocks of black rock, and there were white frothy rapids and longer stretches of clearer water between them. The sides of the canyon were almost sheer in places, and elsewhere dotted with Douglas fir and foothill pine. It was easy to see why the environmentalists were opposed to the dam.

If built, the dam will drown over 10,000 acres of canyon and destroy habitats that support over 200 species of birds, almost 50 species of mammals, and some 30 species of amphibians and reptiles. But there is more to it than that. The canyon attracts over half a million visitors every year. They come to hike, fish, raft, and enjoy nature: the loss of the canyon would be their loss, too. On the other hand, if the American River were to flood, and the Folsom Dam and the levees downstream failed to hold the floodwaters, then there was the possibility that Sacramento would be submerged. In the worst case scenario, hundreds would die, and billions of dollars of property would be lost or damaged—which was why, in the view of Congressman Doolittle and his supporters, the loss of the canyon was a price worth paying.

Midway through my week in and around Auburn, while I was discussing the dam with Eric Peach in his rambling, shaded garden near the gorge's edge, Lowell Jarvis appeared. He had come to collect some pottery fish and he mentioned that he was on the board of the Sacramento Flood Control Agency (SAFCA). He said he could meet me later, if I wished, and he suggested we have a drink at the Fox & Goose, an English-style pub in Sacramento. By the time I met him there I had interviewed many of the key players in the battle over the dam, from engineers at the US Army Corps of Engineers, to the directors of various chambers of commerce; from Congressman Doolittle's chief of staff in Auburn (the congressman was in Washington, DC), to one of the state's leading water attorneys.

Jarvis, a small man with a Chicago accent and a way of talking that made it sound as though he was auditioning for a part in a gangster film, was keen that SAFCA made the right decision on the flood control issue. As an elected representative, he had researched it in depth and he knew everyone I had seen over the past week. "Isn't Ed Tiederman a real gentleman?" he said of the attorney. But the person for whom he professed the greatest admiration was Ron Stork, the senior policy advocate of Friends of the River, a Sacramento-based environmental group. When it came to technical matters, said Jarvis, Ron was the man to listen to. His knowledge about the way in which water behaved, and how the various projects that had been proposed over the past two decades would work, or would not work, was second to none. "Yes," concluded Jarvis, "you can trust Ron Stork."

I think he was right. A man of modest stature, Stork spoke in a quiet voice with great authority, and his round face and spectacles lent him an air of owlish intelligence. "I'm actually pretty formid-able," he said when recalling evidence he had given on the unsuita-bility of the dam to a committee in Washington, DC . "You mean, as a debater?" I asked. "No," he replied, "I wouldn't say that. But I am formidable when it comes to the substance of these issues." He added that many of his opponents, and crucially many of those working for Congressman Doolittle, had a poor grasp of the technical facts.

You could only understand the Auburn Dam issue, suggested Stork, by setting it in the context of the state's long quest to tame its rivers. It was a quest that began in earnest at the turn of the last century when Theodore Roosevelt, a north-east Republican, raised the curtain on the progressive era. California's Central Valley was by then the wheat basket of America, and there had already been some small-scale water control schemes. However, the first wave of big dam building was not conceived until the early 1900s and was executed during the 1920s. Although the schemes made use of federal land, most of the finance was raised locally: in other words, those who were going to benefit from the dam-building paid for it.

Then, beginning in the 1930s, there was a second wave of dam-building, this time on an even bigger scale. "This was Keynsian stuff," explained Stork, referring to the British economist John Maynard Keynes. "You know, big public works projects and the New Deal." During the Depression the state of California was bankrupt, so the federal government, run now by Democrats, paid for the dams, even though it was broke, too. By the 1950s there were major dams on all the Sierra rivers, with the Army Corps of Engineers in charge of flood control schemes and the Bureau of Reclamation in charge of water supply and power projects. In the San Joachim Valley, 98 percent of the river flow is now captured by dams. In the Sacramento Valley, by contrast, there is still some semblance of natural flow on the three forks of the American River upstream of the Folsom Dam, which was built during the 1950s.

These dams—the Folsom, the Shasta, and the Friant, to name but three of the largest—were the pivotal feature of the Central Valley Project, whose purpose was primarily to supply water for agriculture and turn semi-desert into productive farmland. Looked at purely in terms of agricultural productivity, they have been a resounding success. California has the largest agricultural industry of any state. Nevertheless, while farmers—and especially agribusiness —have much to smile about, the taxpayer has less reason to be happy.

The Bureau of Reclamation, established by Roosevelt in 1902, is the federal agency charged with building and operating major irrigation projects, and the relationship between the bureau and the beneficiaries—water agencies and farmers, for example—is laid down by law, which provides for interest-free repayment of construction costs by those who use the water. Between 1902 and 1986, bureau projects cost the taxpayer around US$20 billion at 1986 prices, but so far the beneficiaries have repaid no more than 15 percent of the costs. The Central Valley Project cost taxpayers US$4 billion—it waters over 2.5 million acres on 20,000 or so farms—and farmers often pay as little as US$10 an acre foot for their water (this is the amount of water that covers 1 acre of land to the depth of 1 foot). Studies suggest that even if the bureau factored in a modest interest rate on the costs of construction, and charged accordingly, then water would cost farmers US$60–US$80 an acre foot. If they were to pay the full costs of the water—in other words, sufficient to pay off the capital costs of construction, operation, and maintenance, with interest—they would probably pay double that.

Would this be an outrageous price to pay for a valuable resource? No. Many cities in the West pay in excess of US$300 an acre foot, as do some farmers who get their water from private sources, rather than from federal projects. This suggests that the claim made by farmers that a significant increase in the water fees would put them out of business cannot be taken seriously. Indeed, farmers in California who irrigate their crops with water supplied by state projects, rather than by the bureau, tend to pay four times more than their federally irrigated neighbors.

Quite how good a deal Central Valley farmers are getting was dramatically brought home to me when I later spent time with a group of Hispanic farmers in Colorado. They farmed their land as their ancestors had, without fertilizers or pesticides, and they raised their livestock outdoors. When I asked them what agricultural subsidies they received, they were astonished. None, they replied. But what about your irrigation water, I asked? From where we stood we could see a network of ditches criss-crossing their fields. No, they said, the water was not subsidized. They bought it from a private company and paid around US$600 an acre foot—in other words, 60 times more than many bureau-supplied farmers in California.

One government study suggests that the subsidy to irrigators during 1902–1986 could have amounted to over US$70 billion. Much of the subsidy derives from "interest-forgiveness", but the secretary of the interior can also reduce water charges to irrigators if the interest-free repayment cost exceeds, in his or her view, the irrigator's ability to pay. During the mid-1990s, this led to irrigators in the Central Arizona Project paying just US$2 an acre foot, while the bureau's charges to the local water district were US$60 an acre foot. So this, too, is a subsidy to irrigators, and the munificence of the federal government goes further still: subsidized water is often used to grow crops that are in surplus and eligible for federal subsidies. One government report suggests that this may add a further US$500 million to the subsidy bill each year.

California would not be the California we know today had it not been for this orgy of dam construction; but there has been a high price to pay in terms of environmental damage. Rivers downstream of the dams have been turned into conduits whose water flow is artificially regulated; nutrient-rich sediment is trapped behind dam walls, depriving estuaries of their natural fodder; and as less fresh water flows toward the sea, more salt water intrudes into estuaries, thus bringing about dramatic ecological changes. In addition, considerable acreages of canyon, with their rich mosaic of habitats, have been drowned, and irrigation has enabled farmers to cultivate land that otherwise would have remained pristine.

While nature lovers mourn the taming of wilderness, and the disappearance of species and habitat, thousands of salmon fishermen have suffered a more personal loss. Dams, for them, have meant the loss of their livelihood. Over the past century, 98 percent of the salmon runs on the American East coast have been lost, and today there are more dams in New England—around 3000—than there are wild salmon returning to the rivers each year. Matters are not much better for the salmon on the Pacific coast, at least not in the lower 48 states. According to Zeke Grader of the Pacific Coast Federation of Fisherman's Associations, which represents some 2500 small-vessel operators, the major cause of the collapse of the salmon fishery has been dam-building.

I went to see Grader at his San Francisco office just below Golden Gate Bridge, and found him hunkered down behind a desk so strewn with books and papers that it looked as though he had just been visited by a tornado. The book shelves suggested that he was a man of eclectic tastes. *The History of the American Whale Fishery*, *Black's Law Dictionary*, and various vessel safety manuals and fisheries reports were jumbled up with *14,000 Best Quotes and Quips*, treatises on the English language, a *Rumpole of the Bailey* hat, and a collection of pottery and sculpture.

The modern fishing industry, explained Grader, dated back to the 1850s, when salmon were taken with gill nets in the San Francisco Bay and Delta, largely to provide food for miners. Paradoxically, the fisheries soon suffered from sedimentation caused by hydraulic mining, which sluiced whole hillsides into the Sierra rivers. However, it was the big dams of the 20th century that put paid to most salmon runs by blocking off access to the upper reaches where the fish spawned. "Around 1947 to 1950, the spring-run salmon in the San Joachim Valley went extinct," explained Grader, "and as this had supplied much of the salmon for the gill net fishery, that collapsed, too." The fishery was closed down, and along with the salmon went the gill-netters' jobs. By the 1970s, 95 percent of salmon habitat had been lost. Grader's association estimates that the Pacific north-west salmon fisheries have seen a 90 percent loss of income over the past two decades, primarily as a result of the impact of dam-building, and this has led to the loss of over 46,000 jobs since 1988. The federal and state authorities have tried to compensate for these losses by setting up hatcheries—using, of course, taxpayer money—and these have helped to keep salmon numbers up in some areas. However, hatcheries are no substitute for the real thing: diseases can spread rapidly, there is a loss of genetic diversity, and they are expensive to maintain.

*   *   *

Not long after the completion of the Folsom Dam in the 1950s, the engineers began to draw up plans for the installation of another dam further up the American River at Auburn. However, they were up against what Ron Stork referred to as the diminishing law of returns. While the first wave of big dams was relatively cost effective, the second was not. The Auburn Dam would be twice as large as Folsom, but it would yield only 10–20 percent as much water. Not that this did anything to dissuade the engineers from pressing ahead. Congress authorized the Auburn Dam and construction began in 1967. A Democratic president, Lyndon B. Johnson, signed the papers, and the project received the support of both Democrats and Republicans. There was no significant opposition locally: taming nature, in those days, was seen as a good thing, and the Auburn Dam would certainly do that.

Had it not been for the 1975 earthquake at Oroville Dam some 60 miles away—the earthquake was almost certainly caused by the weight of the dam and its water—there would now be a dam at Auburn. The earthquake sent tremors through the dam-building world, construction ceased, and the engineers set about designing a new earthquake-proof dam. Unfortunately for the dam-builders, the United States now had a president, Ronald Reagan, who was uneasy about the federal government funding major projects that did not yield national benefits. From now on, dams built for the purpose of water storage or hydropower would have to be funded locally. Dams built for flood control purposes would receive 75 percent federal funding, with local interests providing the rest. The cost share is now set at 65 percent federal, 35 percent local. All the same, this still means that "pork", as Stork called it—in other words, taxpayer dollars—is available for flood control projects.

The supporters of a dam at Auburn were forced to change tack. They no longer argued for a water storage facility; if they had, they would have needed to find the funds locally. Instead, they argued that a dam was necessary to control flooding on the American River. Fortunately for them, there had recently been several high water events, and these had concentrated the minds of those who lived in the flood plain.

At the time of my visit, everybody agreed that the Folsom Dam, as it stood, and the levees that channeled water past Sacramento, provided insufficient protection. A major flood on the American River could devastate the capital. There was much disagreement, however, on precisely what measures were required to prevent flooding. The environmentalists favored the latest SAFCA scheme, which would modify the Folsom Dam and raise the levees downstream. Congressman Doolittle, many business interests in Sacramento, and the local staff at the Army Corps of Engineers, were adamant that the highest levels of protection could only be assured by the building of an Auburn Dam. Anything less, they said, and there would be an unacceptably high risk of flooding.

This was not just about the risk of flooding—and on this subject the two sides were unable to agree how much protection is required —it was also about money. How much could justifiably be spent and whose money should it be? "Money is the big problem," suggested Bruce Cosgrove, a former chairman of the Auburn Dam Council and now director of the Auburn Chamber of Commerce. "There's this underlying belief that someone else ought to pay for it. If you've grown up in a world where the federal government pays for [big projects], why would you raise your hand and say: 'Put me down for US$200 million'?" You wouldn't.

Both sides accused the other of behaving in a cavalier manner towards taxpayer money. "The environmentalists always come out as fiscal hawks when it comes to the Auburn Dam," said Richard Robinson, who worked for Congressman Doolittle and looked like a fiscal hawk: young, lean, impressively tall, and immaculately groomed. "But when the levee plan is debated, the sky is the limit. We would contend that the enviros don't really care much about flood control. They care about killing the Auburn Dam and that's their one and only objective, and the levee plan is what does that."

Everyone seemed to agree that if Congress were to provide funds for Folsom modifications and levee raising, which might cost around US$600 million, it certainly would not countenance the spending of a further US$1 billion on a dam at a later date. The former would effectively kill off the latter. The environmentalists, for their part, pointed out that their favored scheme might be expensive, but it came in at half the cost of the dam. They were doing the taxpayer a favor, as well as the people of Sacramento, who would not have to pay so much as they would if the dam was built. In contrast, Congressman Doolittle and his supporters were not really interested in getting a good deal for the taxpayer. Look, said the environmentalists, at the games he played in Washington, DC, over the past decade.

In 1992 Congressman Doolittle opposed a flood control dam at Auburn, siding with the majority of politicians in the House of Representatives who, for either fiscal or environmental reasons, voted against the dam, which at that time was supported by SAFCA and the Army Corps of Engineers. According to Ron Stork, Doolittle opposed the dam not because he was concerned about taxpayer money, but because he feared that the proposed dam would not be expandable. His dream was to have a dam that would provide water and power in the future, and thus help foster development in his district.

The shock of defeat in 1992 was so great that the staff of the local Army Corps of Engineers were given grief counseling. "They went into hiding," recalled Stork with a wry smile. "We didn't see that for a year." When they eventually re-emerged, the corps reluctantly agreed to cost out alternative schemes, including those that would involve modifications to Folsom and the existing levees. In 1996 SAFCA studied the latest flood control options put before it by the corps, and its members voted narrowly in favor of requesting federal funding once again for the Auburn Dam. Significantly, most of those representing the flood plain, whose inhabitants would be taxed to pay for the local share of the dam, voted against. Those individuals whose constituents lived outside of the flood plain, and who were not at risk of flooding or extra taxes, voted in favor of the scheme. By nature, human beings are generally far more generous with other people's money than they are with their own.

"You never saw a happier colonel [of the corps]," said Stork. "He was relieved his baby had come home." So, too, was Congress-man Doolittle, whose district would play host to the dam and benefit economically from its construction, yet pay nothing towards the cost, as the local share would be paid mainly by those living in the flood plain. However, the Auburn Dam scheme was again the subject of debate in Washington in 1996 and 1998, and again it failed to get the support required.

Long after I left California, I received a message from Charlie Casey, a colleague of Ron Stork at Friends of the River. The Auburn Dam project, he wrote, was effectively dead. The authorities had decided to solve Sacramento's flood problem by going down the route suggested by the environmentalists, by raising Folsom Dam and by improving the levees. "However," he continued, "you may have heard about the recent energy crunch and the problems related to power deregulation and California." Yes, I had. Botched attempts to deregulate the electricity market some five years ago had led to the two main utilities buying power on the expensive wholesale spot market, and selling it at the lower rate to consumers. This was a recipe for disaster, and the state was plagued by black-outs and faced serious price hikes. All of this was good news for those individuals—among them Congressmen Doolittle—who were looking for a reason to resurrect the Auburn Dam scheme. "Just yesterday," wrote Casey, "I received a copy of his [Doolittle's] latest editorial in the local paper in which he calls for the immediate construction of the Auburn Dam, and essentially blaming all of society's ills on the environmentalists."

Let us return, briefly, to the subject of money—in particular, the money of American taxpayers. Why would a politician back such an expensive scheme, a dam at Auburn, when a cheaper, less destructive option was available? The most obvious explanation for Doolittle's obsession with the dam is that it would channel large dollops of federal money into his district and foster a development boom. As Bruce Cosgrove said when I saw him: "It's great politics. It's wonderful for Mr. Doolittle or any congressman in any state to [obtain] federal funds for any common good interest, whether it's a dam or a rail system"—regardless, one might add, of whether it's needed or not. Politics is about bringing home the bacon or, as the Cato Institute libertarians would have it, looting the national taxpayer to enrich your local community.

It seems, too, that there is a religious element to this love of big projects. Ron Stork made much of the fact that Congressman Doolittle was a Mormon. The Mormons, and especially those who subscribe to the so-called "dominance paradigm", believe that nature must be bent to man's will. This is also a view espoused by many Christian churches. On several occasions I heard politicians quote a verse from the prophet Isaiah as justification for the building of dams, highways, and other major projects:

> Prepare ye the way of the Lord, make straight in the desert a highway for our God. Every valley shall be exalted, and every mountain and hill shall be made low: and the crooked shall be made straight, and the rough places plain.

Translate this into the modern idiom and you get: God is on our side, so let's do it. Let's build the Auburn Dam, carve a highway through this mountain range, irrigate these deserts.

If you happen to live in a part of the United States that never floods, you might take exception to the idea of your tax dollars paying for flood control schemes in places such as Sacramento. In effect, you are paying to ensure that home-owners elsewhere are protected from flooding. So why, you might reasonably ask, is the federal government paying over two-thirds of the bill for projects that yield no benefits for those who do not live in the flood zone? The stock reply is that if cities such as Sacramento ever flood, it will cost the federal government billions of dollars in emergency relief; so it is cheaper to pay for a flood control scheme now, than for a clean-up later. This is certainly a tricky issue, for every state at some time draws upon federal money for its own, rather than the national good.

To some extent, the federal government only has itself to blame, as decades of largesse have helped to create flood problems in the first place. In the days when those individuals and communities who were hit by floods had to clean up the mess at their own expense, and received nothing by way of compensation, only the foolish built their homes, or established their factories or farms, in places where there was an obvious risk of flooding. Flood insurance and emergency relief programs changed all of this. Millions relocated on flood plains, and as their numbers rose, more dams and levees were built to control water, at the taxpayers' expense. Between August 1995 and June 1998, the National Flood Insurance Program, set up by Congress under legislation passed in 1968 and 1973, had a net borrowing from the Treasury of US$810 million. Astonishingly, almost a half of the money paid out under the National Flood Insurance Program has gone to the owners of "repetitive loss properties", who make up just 2 percent of all policy-holders. This means that the taxpayer is subsidizing the imprudence of a few.

The taxpayer has also been taken for an exceptionally expensive ride on the 11,000-mile long inland waterway system. According to the Congressional Budget Office, this is by far the most heavily subsidized commercial freight transportation system in the United States. Prior to 1981, barge operators paid nothing towards the operation and maintenance of the system, and nothing towards the cost of constructing the canals, locks, and so forth. This is another rich man's racket, with just 20 companies—including such big names as Cargill and Continental Grain—owning over four-fifths of all the barges operating on the Gulf Coast Intercoastal and Extended Mississippi system, which makes up most of the inland waterways. In 1998, the US Army Corps of Engineers spent US$766 million on the inland waterways; industry paid a mere US$79 million of this. Although a tax on barge diesel fuel yields revenues that cover 50 percent of the costs of new construction, taxpayers still pay for 100 percent of the operation and maintenance cost and 90 percent of all inland water costs, including half of new construction.

It is worth reflecting, for a moment, on the way in which transport subsidies, such as those that help enrich the corporations mentioned above, distort the market in agricultural products. Frequently, I was told that a free trade in foodstuffs—in other words, in unsubsidized produce that is not subject to tariffs, quotas, or other measures that hinder its sale—would lead to the loss of local markets. If, say, Argentina can produce beef and transport it to Switzerland or New Mexico more cheaply than farmers there can produce beef, then the Argentinians will win. If, say, Australia can produce lamb and transport it to Montana more cheaply than farmers in Montana can rear lamb, then the latter will lose out as consumers in Bozeman opt for the best deal they can get.

This might indeed be the case, but that is partly because transport —by road, air, or barge—is so heavily subsidized. If hauliers paid the true cost of using US interstate highways and freeways, and supermarkets could not take advantage of a range of indirect transport subsidies, then the whole system of food distribution would be more localized. Farmers who are currently disadvantaged by transport subsidies would be much better able to compete with "cheap" imports. As the British philosopher and *Financial Times* columnist Roger Scruton suggests:

Once you establish distribution systems which reach out indefinitely and which are also publicly maintained, you abolish the free economy forever, and condemn all those who cannot externalize their costs—among whom small farmers are most prominent—to extinction.

All too often, one subsidy leads to, or necessitates, another. Just look at what is happening in the American West with its rivers. Every year the US taxpayer forks out around US$170 million for a salmon restoration project on the Lower Snake River. The US Army Corps spends this considerable sum of money transporting salmon around the four major dams in trucks and barges. In 1999 it was planning to spend a further US$425 million on new capital construction activities because the existing scheme was failing to save the salmon. According to environmentalists, it has been a disaster, and the 1998 return rate for fish barged in 1995 was a quarter the level required to prevent extinction. It would be cheaper and more effective to partially dismantle the four Lower Snake River dams. If this ever happens, it will mean that taxpayer money has been used to construct dams on the Snake River; to finance projects on the Snake River that have enabled multinational companies to ship subsidized grain around the states; to pay for the transport of salmon whose survival has been threatened by the above; and, finally, to pay for dismantling the structures that were financed by an earlier generation of taxpayers. It is a curious way for a government to treat its citizens.

* * *

When it comes to wasting and abusing water, the global picture is grim. The failure to make water users pay the true cost of water supply has encouraged farmers, industries, and domestic consumers, to treat it as though it were virtually free, and make little attempt to use it sparingly.

In 1998, for the first time in history, the number of people displaced by environmental problems of one sort or another exceeded the number (21 million worldwide) displaced by war.

According to the World Commission on Water, the number of environmental refugees could quadruple to 100 million over the next 25 years. The abuse of water is one of the key factors behind this great exodus, and many of the world's largest rivers and watersheds are now in trouble—trouble caused, partly or wholly, by the underpricing of water, and the subsidizing of its provision.

The flow of the River Ganges, which channels the melt water of the Himalayan mountains through India into the Bay of Bengal, has been so severely reduced by extraction that the delta swamps of the Sunderbans, home to the Bengal tiger, are becoming progressively drier. Only 3 percent of the Ganges' water is safe to drink, such is the pollution from domestic waste and industry. The Yellow River in China is likewise severely polluted and overexploited. During 1997 its lower reaches were completely dry for over 200 days. In Africa, the Nile loses 90 percent of its water to irrigation and other uses, and the river is severely polluted. In the Middle East, the River Jordan is so heavily used that only one third of its water makes it to the Dead Sea.

Perhaps the most famous example of river abuse comes from Central Asia. Less than half a century ago, the Amu Darya and Syr Darya rivers delivered 55 billion cubic meters of water each year into the Aral Sea, which in those days was the Earth's fourth largest lake. The Soviet Union then initiated a scheme during the 1960s to transform the neighboring desert into a vast cotton estate. To do so it bled the rivers almost dry, and by the 1980s only 7 billion cubic metres of water reached the Aral Sea each year. As a result, the fisheries which once supported 60,000 people have collapsed, two dozen native fish have become extinct, and the population of Muynak, a once prosperous fishing town, has fallen from 40,000 to just over 10,000. The abuse of the Aral Sea was fueled by subsidies.

Almost everywhere the propensity of governments to undervalue water, and subsidize its use, is leading to both its waste and to severe environmental problems. In Bangladesh, farmers pay 1 percent of what it costs the government to supply them with water; in other words, there is a 99 percent subsidy. In the Philippines, farmers pay 10 percent of the cost of supply; in Mexico, 11 percent; in Egypt, 20 percent. The cost of building the infrastructure that supplies the water is not even factored into these calculations. If it were, the subsidies would be even greater. I am wary of the figures that are banded about for such things as irrigation subsidies because governments seldom keep good records; but there is no reason to believe that the figures provided by the World Bank are far off the mark. The bank estimates Africa's annual irrigation subsidies to be around US$6 billion; Latin America, US$3.4 billion; and Asia, US$12.5 billion.

In the developing world, governments often justify water subsidies by arguing that they help the poor, whether they are poor farmers or the impoverished inhabitants of big cities. In rare instances, they may do precisely that. Mostly, they do not. As far as domestic consumers are concerned, the poor often pay far more for their water in developing world cities than the rich. While the latter have access to municipal water supplies that are invariably subsidized, the poor and those who live in squatter settlements are forced to pay many times more for their water, which they purchase from private water vendors. Take Port-au-Prince, Haiti. Here slum dwellers are paying 100 times more for their water than the self-styled "morally repugnant elite", who get municipal water supplied direct to their mansions.

If water subsidies were phased out, what effect would this have on the way we use water? Without doubt, farmers, industries, and domestic users would seek to reduce their consumption. Encouragingly, Israeli farmers, who are obliged to pay much more for their water than most, now get five times more in value for their crops per unit of irrigation water than they did in the past, when water was cheaper. This has been achieved by replacing old-fashioned irrigation systems, which are inefficient, with drip irrigation, which is far less wasteful. Pioneering work on the creation of water markets in the United States, in areas where water is treated as a salable commodity rather than as a free resource to be frittered away, is encouraging farmers to conserve their water in order to sell it to other users. Municipalities and industries have also shown that through recycling and other measures, water consumption can be reduced dramatically.

Back, briefly, to California, a state whose projected population growth is as spectacular as that for the most fecund and contraceptive-free region in India. Today the population is around 33 million, and is projected to rise to 50 million by the year 2020. Supplying the population with adequate water is going to be a major challenge. "Water conservation is going to be the big issue," suggested Lowell Jarvis in the Fox & Goose. "At present, folks in southern California use 250 gallons a day each. Up here we use 500 gallons, and we're going to have to reduce that." By this he meant that all forms of water use—agricultural, industrial, domestic—accounted for 250 gallons per capita in the south, double in the north. The environmentalists agreed that conserving water had to be the priority if the state was not to run dry. Charlie Casey told me that until he was 42 years old—he wasn't much older now—he had never seen a water meter. A mere 40 individuals in Sacramento, Casey being one of them, had water meters now; but because consumers paid a flat rate for their water there was no incentive to reduce waste. In one water district nearby, there were 165 miles of open ditches delivering water—this in a climate not dissimilar to the Sahara, with very high rates of evaporation. This was absolute madness, suggested Casey. Add to this the wastage by farmers and industries and clearly there was considerable scope for reducing water needs.

Zeke Grader was not so convinced. "The enviros have been seduced into looking at technical fixes," he said, "instead of saying, 'Whoa! Let's step back a moment, folks, and take a look at this.' Even if water consumption is going down per capita, we're still going to have more demand on water supplies. Where will it come from?" Congressman Doolittle's supporters would say to this: from projects such as the Auburn Dam. But there's a problem. If built, the dam would increase California's stock of captured water by 0.05 percent, according to Lowell Jarvis. This brings us back to conservation: the best way to meet future demands is to use less. If that is to happen, water must be priced as the valuable resource that it is.

Unfortunately, there are plenty of people in power who are not remotely interested in using resources more frugally. When the Plumbing Standards Act 2000, otherwise known as the Potty Law, was being debated—its aim was to do away with certain regulations encouraging efficient use of water—an aide of Joe Knollerbury, a leading politician proposing the new law, memorably told the press, "We are used to getting a good shower in the US of A and, by God, that right has been taken away from us by bureaucrats in Washington." This puts me in mind of a remark made by Mark Twain. "Suppose you were an idiot," he said, "and suppose you were a member of Congress. But I repeat myself."

CHAPTER 5

NATURE: ON THE WANTED LIST

I recently stumbled across a book called *The Last Stand of the Pack*, written, I think, during the 1930s by Arthur A. Carhart and Stanley P. Young. Young was principal biologist at the US Biological Survey, which oversaw the government program to control predators such as the wolf. The following paragraph sums up the tenor of the book, which describes the hunting down of Colorado's last wolves:

> Man has won. The wilderness killers have lost. They have written their own death warrant in killing, torture, blood lust, almost fiendish cruelty. Civilization of the white man has almost covered the West. And with that nearly accomplished, there was no place left for the gray killers, the renegades of the range lands.

Prior to the European arrival five centuries ago, there may have been as many as 400,000 gray and red wolves in North America, but by the 1930s they had been exterminated almost everywhere in the lower 48 states apart from in Minnesota, where less than 1000 survived. The wolves were wiped out primarily because ranchers, and the authorities, saw them as a threat to the livestock industry: it was a question of us or them. Recently, however, there has been a great change in attitude towards predators, both among the public and within certain sectors of government, and over the past decade federally funded wolf reintroduction programs have enabled small populations to flourish and spread back into some of the habitats their ancestors once occupied.

The story of the wolf is in many ways the story of the American West. For centuries nature was something to be conquered. The "civilization of the white man" meant the killing, or subjugation, of everything that stood in his way, whether they were wild animals or Indians. But now, Americans increasingly see nature as something with which they should coexist peaceably, or even revere.

When Alexis de Tocqueville traveled around America during the 1830s, he contrasted the European attitude towards wilderness— nature was by then much admired by the romantic poets and the public—with that of the Americans. "They are insensible to the wonders of inanimate nature," he wrote, "and they may be said not to perceive the mighty forests that surround them till they fall beneath the hatchet." What they were really interested in, noted de Tocqueville, was "draining swamps, turning the course of rivers, peopling solitudes, and subduing nature."

Gradually, siren voices alerted Americans to the non-material values of the world that they were so busy subduing. Henry David Thoreau, the author of *Walden*, announced that he wanted "to speak a word for nature, for absolute freedom and wildness." For transcendentalists such as Thoreau, wilderness was a place where man could find his spiritual self. Later, John Muir, who was to become the first president of the Sierra Club, noted that: "The universe would be incomplete without man; but it would also be incomplete without the smallest trans-microscopic creature that dwells beyond our conceitful eyes and knowledge." Men such as these shaped the thinking of Aldo Leopold, Barry Lopez, Richard Nelson, and a host of others whose writings are now far more influential than the sort of melodramatic tosh that I quoted at the start of this chapter.

Not that all is peace and light. There is still bitter conflict between those who believe that the government should be spending taxpayer dollars on controlling predators that threaten livestock, and those who don't. And there is still bitter conflict between those who believe that the government should be spending money on conserving nature and protecting endangered species, and those who don't. That is what this chapter is about.

*   *   *

For centuries the killing of predators in the American West was a freelance affair, left to trappers, bounty hunters, and ranchers; however, in the early years of the 20th century, the federal government decided to lend a hand. The president then was Theodore Roosevelt, sometime rancher and an ardent hunter who saw wolves as "cruel, crafty beasts". In one of its first years, the department that was established to control predators killed 1800 wolves, 23,000 coyotes, and many hundreds of lynx, mountain lion, bear, and bobcat. The program even received the support and backing of conservationists. In those days, Yellowstone National Park was run by military superintendents who treated all predators as enemies. Grizzly bears were blown up with dynamite as they scavenged around garbage dumps; wolves, coyotes, and bobcat were trapped, shot, and poisoned; and a pack of hounds was acquired to hunt down mountain lions. Lynx, marten, otter, and weasel were also killed in the quest to make Yellowstone a Garden of Eden for elk, moose, and other edible game. However, as the years slipped by a new breed of ecologists began to realize that predators played an important role in maintaining healthy ecosystems, and during the 1930s they convinced the park authorities to abandon the predator eradication program.

But the killing has continued outside of protected areas, with the blessing of every administration. Year after year, the US taxpayer has footed the bill for this lethal control program, which is now run by the US Department of Agriculture's (USDA's) Wildlife

Services. During the financial year of 1997, some US$10 million was spent on killing (and this is just a sample) 321 black bears, 1845 bobcats, 82,392 coyotes, 3683 red foxes, 320 mountain lions, and 212 gray wolves. Many other species were targeted, too. The latest available figures, for 1999, show that Wildlife Services killed over 96,000 predators.

I began my investigations into the predator control program in Helena, Montana's sleepy little capital, and my first port of call was the cramped office of the Montana Wool Growers Association. Its secretary, Bob Gilbert, had hardly finished shaking my hand when he announced, through gritted teeth, that coyotes were public enemy number one in the West. A stocky individual with wiry gray hair, a florid complexion, and a truculent manner, Gilbert shared his chair with an incongruously pretty Shih Tzu dog whose plumes of facial hair were primly secured with purple bows.

It was easy to understand why Gilbert was angry. Over the past few years wool and lamb prices had plummeted, largely because there was a worldwide glut, and matters for Montana's 1800 or so sheep farmers had been made worse by the aggressive marketing of cheap imports from New Zealand and Australia. The sheep farmers had seen their income drop by one third, and the secretary of the Wool Growers Association had recently taken a US$20,000 salary cut. Then there was another menace, in the shape of the animal rights campaigners who wanted to close down the government's predator control program. For his members, he said, the program was an absolute necessity. In 1998, predators caused US$1.1 million in losses to the sheep industry in Montana, accounting for 33 percent of all deaths. Coyotes were the prime offenders, killing 14,900 sheep and lambs, or 68 percent of the livestock taken by wild animals; but black bear, mountain lion and bobcat also caused problems.

Later that day I crossed town to the offices of the Montana Stockgrowers Association, where I met Beth Emter, the handsome, silky-voiced spokeswoman for the state's beef industry. "Predators can make or break a business," she suggested, and one way or another she reckoned they affected most of the state's ranches. On her parents' ranch, the previous year's calving rate had been reduced by one tenth as a result of coyote attacks. "The government predator-control program is absolutely critical for ranchers," she said, adding that there was no question of ranchers wanting to exterminate the coyote or any other wild animal. However, they had to be kept under control, otherwise many ranchers would go out of business.

After I left Beth Emter, I headed south to visit the Farm Bureau in Bozeman, a small university town situated in a bowl of land to the south of the Gallatin valley and a little way north of Yellowstone National Park. Montana may be best known to outsiders as a rugged cowboy state, but it also has great swathes of arable land, and much of the journey took me through rolling wheat fields that stretched from one horizon to the other. However, the further south I journeyed the more varied the landscape became, and the wheat fields eventually gave way to green pastures with grazing cattle and sheep.

I already had an inkling of what I would hear at the Farm Bureau, an organization with little sympathy for environmental matters. Three years ago the Farm Bureau had gone to court in an attempt to get the wolves that had been re-introduced by the government to Yellowstone National Park and Idaho either killed off, or sent back to where they came from, which was Canada. The Farm Bureau won the initial court case; but the case had now gone to appeal and the decision was expected to go against them. However, my conversation with John Youngberg, one of the bureau's directors in Bozeman, began not with talk of wolves or coyotes, but of the behavior of the packing industry, against whose quasi-monopolistic practices Youngberg railed. Once we had exhausted that subject we turned to the business of predator control, and Youngberg was full of praise for the work of Wildlife Services.

But why, I inquired, should taxpayers from, say, Cleveland, Ohio, or Jackson, Mississippi, pick up the bill for killing some 10,000 predators each year in Montana? He gave the matter some thought, then said: "When I was young, everybody's grandpa had something to do with agriculture. But now things have changed so much, and most people are far removed from the land. That means they don't understand the problems." This is undoubtedly true. During the 1860s five out every six Americans lived in rural areas. Now less than 2 percent of the population of the United States are directly employed in agriculture, and the majority of Americans have little or no knowledge about where their groceries come from, or about how their crops are grown or livestock reared. A predominantly ignorant public, suggested Youngberg, failed to understand how predators could threaten livestock farming; yet they happily swallowed the propaganda of animal rights and environmental groups who opposed the government control program.

"What we're seeing," argued Youngberg, "is a small proportion of the population paying for what the rest of society wants." By this he meant that ranchers were paying the penalty for doing business in areas where coyotes and other predators, much loved by society, thrived. Logically, therefore, society, in the shape of the taxpayer, should compensate them for this state of affairs by paying for the predators to be kept under some sort of control. Both he and Gilbert suggested that because many of the predators came off public land to kill privately owned livestock, the public was under moral obligation to contribute towards the cost of predator control.

Bob Gilbert was outspoken in his criticism of the organizations who were campaigning against the predator control program. One of the most influential of these was the Bozeman-based Predator Conservation Alliance, and Gilbert was robustly contemptuous of its director, Tom Skeele. "It's interesting," said Gilbert wearily, "that the people who don't want predator control come from a vegetarian standpoint. They're trying to make sure others don't eat meat either." The Predator Conservation Alliance was an animal rights organization, he said, and Skeele was a vegetarian. He expelled the word vegetarian from his mouth as though it was a chunk of rancid meat.

When I repeated Gilbert's remarks to Tom Skeele, a balding, athletically built East Coaster—this is also a term of abuse among many Montana natives—he said: "We are a conservation organization, not an animal rights organization." He then expanded on this: "A lot of the work we're doing is based on the notion that all animals, including predators, have a right to exist for their own being." This sounded to me like a good definition of an animal rights organization, but he added that they were not campaigning against Wildlife Services just for the sake of individual animals, as an animal rights organization would, but for the sake of the whole natural system.

The Predator Project, the forerunner of the Predator Conservation Alliance, was set up by Skeele in 1991. Since then it has produced a steady stream of publications analyzing the work of Wildlife Services, or Animal Damage Control as it was more honestly known until recently, and refuting what it sees as the lies peddled by the government agency. Armed with one of its latest publications, *Ten Myths that Perpetuate Environmental Destruction and Government Waste*, and with some of the website literature of Wildlife Services, I was able to put similar questions both to Skeele and to Wildlife Services' Montana director, Larry Handegard, whom I later met at his office at Billings airport—a sensible enough location in view of the fact that much of the predator control is done from helicopters and fixed-wing aircraft.

Wildlife Services and its defenders have always claimed that it provides broad benefits to the public. Whenever the agency's predator control program has come under threat on Capitol Hill—as it did during both 1998 and 1999—its supporters have played this card. "Ask yourself," said Congressman Henry Bonilla, a Republican from Texas, "if you could live with an accident occurring at an airport—or live with a child dying, who was affected with rabies—because there was not enough money in the budget to support this program." When I asked Handegard what would happen if Wildlife Services' budget was cut, he used a similar argument.

Sure enough, Wildlife Services is involved in preventing bird strikes at airports and in various measures that protect human health. However, half of its budget is channeled into predator control programs in the 17 Western states, while programs that benefit the average American citizen account for just over one tenth of its budget. In Montana, for the financial year of 1999, Wildlife Services' total budget was US$2,049,897, of which the federal government provided US$928,091, over 90 percent of which was spent on livestock protection—in other words, on killing and controlling wild animals. In any case, Congressman Bonilla's tear-jerker of a speech was thoroughly disingenuous: nobody ever suggested that funds for such things as airfield bird control or rabies prevention should be cut.

Handegard pointed out that livestock farmers and the states contributed towards the predator control program, providing around half of the funding in Montana. Skeele was unimpressed by these computations, and he claimed that in the West private ranchers directly contributed no more than 1 percent of the costs. On reflection, I think both were putting a misleading gloss on the figures. The fact is that in Montana during, for example, 1997, the federal government provided 57 percent of the funds, the state provided 21 percent, and farming organizations provided 20 percent. If we take the 17 Western states together, the federal government provided 49 percent of the costs of predator control; the states, 24 percent; the counties, 11 percent; and farming organizations, 15 percent. Crudely put, there is a 75 percent subsidy, two-thirds of which comes from Washington, DC.

When I saw Skeele, he made much of the fact that the cost of Wildlife Services' operations in the West exceeded by a factor of three the value of livestock lost to predators. Therefore, he argued, the program made no sense financially. However, Handegard rightly pointed out that this was a nonsensical argument. What matters is what the losses would be if Wildlife Services were not killing predators. If they exceeded in value the cost of the program, then in purely financial terms the program would make economic sense —providing, of course, that it was entirely funded by the beneficiaries, and not by the taxpayer, which it isn't.

But does the program actually work? Or, put another way, is taxpayer money achieving the ends for which it is intended?

To get a rancher's point of view, I went to see John Paugh, a retired sheep farmer who now lived in a sub-division near Bozeman. A small man with arthritic hands, gray hair, and spectacles, Paugh was a likable character, much given to reminiscing about the past. "I know just how the Indians must have felt when the white men came here," he said with a chuckle. "'Here are all these damn people messing up our lives! They're changing the way we live, and how we do things.'" Over the last 20 years the population of his county had doubled, and droves of newcomers had pushed up the price of land, clogged up country lanes with their U-haul trucks and sports utility vehicles, and at times let their dogs run loose to kill livestock. And among the ranks of the newcomers were many environmentalists who were doing their best to make an already difficult life even more difficult.

Paugh was emphatically in favor of Wildlife Services, whom he and his sons had frequently called upon to help control predators and protect his family's 1500 sheep. Without the predator program, he said, the losses would be too great to bear. In fact, Paugh harked back to the happier days when the poison 1080, now banned in the United States for most purposes, killed predators far more effectively than the present methods of shooting from aircraft and trapping.

On the contrary, said Skeele, the program was not working. Research suggested that when coyote populations were stressed— by, for example, shooting and trapping—the survivors reacted by producing larger litters. Furthermore, as soon as the animals were killed in one area, others moved in from elsewhere to take their place. According to the Predator Conservation Alliance, the number of coyotes killed by government trappers and shooters in Montana doubled from 4530 in 1987 to 8720 in 1995; yet the number of calves killed by predators had remained constant since 1991. Handegard disputed the research that suggested that killing led to an increase in litter size, and he also pointed out that coyote numbers had risen dramatically because there were fewer private trappers than in the past, the bottom having fallen out of the fur market.

If you log onto the various USDA websites that cover the issue of predator control—there is even one aimed at school children— you will find vast quantities of information, but not, strangely, any mention (or none that I could find) of precisely how many ranchers and farmers use Wildlife Services to control predators. So I asked Handegard how many ranchers his 19 field officers helped each year. He said he didn't know; there were no figures for this. But he said that the number seeking help was rising, largely as a result of an increase in coyote predation on calves. When I asked by how much it had increased, he replied: "I don't have a number." Frankly, I find it hard to believe that an organization who can give a budget breakdown to the last dollar, and exact kill numbers of everything from coyotes to porcupines, fails to keep records about the number of ranchers it assists. And the only reason I can see for not revealing the figures would to be conceal the fact that a great many ranchers do not benefit from Wildlife Services.

Indeed, some ranchers, such as Becky Weed, are openly critical of the predator control program. I headed out to see her one pleasant afternoon as fluffy white clouds skidded across a brilliant blue sky. Weed lived at Thirteen Mile Farm in the Gallatin Valley, and the short journey from Bozeman took me through some lovely countryside. It was now harvest time. Some fields of wheat had already been cut; others awaited the harvester. Sleek cattle grazed in lush meadows and on either side of the road the fields were sprinkled with stately wooden barns and rambling homesteads.

A blond-haired woman with luminous blue eyes, Weed suggested that we go on a tour of her small farm, and she led the way past four huge cottonwood trees and a fine old barn into a patchwork of small meadows. She explained that she and her partner had bought the 160-acre farm in 1987, and she soon became obsessed with learning how to manage the land decently. She now had a flock of 180 ewes, and she direct-marketed her lamb and sold her wool through the Growers' Wool Cooperative, of which she was president. Her flock was certified as "predator friendly" by Predator Friendly Inc., which numbered Tom Skeele among its directors. To be certified, ranchers pledged to protect their livestock using only non-lethal methods of predator control.

"There's definitely a problem with predators," said Weed as we passed a batch of lambs that were due to go for slaughter, "and in some places it can be catastrophic. But we've found that shooting coyotes doesn't work. The coyote population seems to be going up and predation isn't going down. Transients move in and litter sizes seem to increase when you stress populations." She realized soon after she moved into the sheep business that the predator control programs were also antagonizing local environmentalists, and that, in any case, predators had a role to play. Among other things, they controlled pests such as gophers and rats and they disposed of carrion.

Weed decided to look at other ways of limiting predator kills and she acquired a lama in 1994. "It's been very effective," she explained. "Lamas are territorial like coyotes, and if coyotes approach the sheep our lama will kick at them and chase them off." Weed had also changed her pasture management to deter predators. Two years ago a black bear had got in among her sheep in one of the more distant meadows and killed a few ewes, so she moved the sheep closer to her house and eventually the bear left. She had also had a problem this spring with a bald eagle that preyed on some of her early lambs. However, as soon as the gophers came out of hibernation, the eagle preyed on them. "If you do something unnatural, like lambing early," said Weed, "then that's the sort of thing that happens."

It says much about the attitude of many ranchers towards predators that when the predator-friendly initiative was launched, farmers' organizations and the agricultural press ridiculed the idea, and Weed received hate mail and threatening phone calls. One large rancher who joined the Growers' Wool Cooperative in the early years later left because his stockmen disliked the idea and his peers shunned him. He even had trouble finding a conventional market for his wool after quitting, as the wool merchants wanted to punish him for what they saw as an act of betrayal.

The cooperative had now been in operation for four years and it was clearly here to stay. But Bob Gilbert was dismissive; he suggested that the only people practicing predator-friendly sheep ranching were small operators who had other sources of off-farm income. This was also the line that Paugh took. It was all very well for Weed to use non-lethal methods of control, he said, but she had little land, few sheep, and a partner who made a living elsewhere. Over the years Paugh had tried all of the available non-lethal methods of control, using guard animals, burros, and the like. None provided sufficient protection for a rancher who was running a large band of sheep over thousands of acres of public land. The Paughs still used guard dogs, but they also shot, hunted, and called in Wildlife Services to control predators. Paugh paid to graze his sheep on public land, and the federal authorities, in his view, had an obligation to make sure that his operation was viable; without predator control, it would not be.

Here I must declare a prejudice. I do not believe in animal rights, and I am profoundly disturbed by the often misanthropic attitude of those who do. I see nothing wrong, either morally or ecologically, with the killing of wild animals, providing it is done as humanely as possible, and in such as a way as not to threaten the survival of the species. However, nothing I heard or saw convinced me that the US taxpayer should pay to control predators. Those ranchers who use Wildlife Services would be justifiably angry if they discovered that a slice of their taxes was being used to sort out death-watch beetle infestation in a warehouse in Boston, or for the killing of rats in a restaurant in Manhattan, or as compensation for the death of a Shih Tzu, devoured by a coyote in the suburbs of San Diego. It is equally unreasonable to expect taxpayers to pay for the cost of doing business on sheep and cattle ranches in the West. Predators are a feature of the world in which ranchers operate; and, indeed, many realize this. "He's part of our West," said the cowboy poet Wally McRae, referring to the coyote. "And I don't think you can eliminate by fiat things that don't suit you."

I accept the argument put forward by ranchers such as Paugh that non-lethal methods of control are not sufficient to protect large bands of sheep grazing great swathes of public land. But that is the ranchers' problem, not the taxpayers'. The truth is that Montana's public lands are not a great place on which to graze sheep. As Aldo Leopold once said, sheep are animals that are looking for a way to die.

All of which begs the question: how can such a controversial and perverse subsidy be persistently approved by Congress? In June 1998 the House of Representatives voted 229 to 193 to cut US$10 million from Wildlife Services' predator control program; but the following day there was a revote, which went 232 to 192 in favor of continuing the subsidy. In 1999 an amendment proposed a US$7 million cut in the appropriations to Wildlife Services. On July 23, it was approved by 229 to 193; but the next day it, too, was subject to a revote, and was defeated by 230 to 193. The reason why opponents of Wildlife Services' predator program chose to go for an amendment, rather than seek change with a provision from within the sub-committee on agricultural appropriations, was because the program had the support of Joe Skeen, the Republican committee chairman. Skeen regularly used Wildlife Services to control predators on his ranch in New Mexico. It was Skeen who invoked an obscure parliamentary procedure that led to the revote. Over 30 representatives changed their minds within a period of 24 hours. They had come under intense pressure from the Farm Bureau and its supporters, who were aided by a statement from the secretary of agriculture, Dan Glickman. Glickman said that if the amendment passed, he would shift the funding cuts to those parts of the program—

rabies control, protection of airports from bird strikes, and so on—
that actually benefit the public. This sounds like blackmail to me.

Since 1997, the annual appropriations for Wildlife Services have
increased dramatically. In 1997, the USDA requested US$26,642,000
in federal appropriations; Congress agreed to US$26,947. For the
financial year 2001, the USDA requested US$28,864,000; Congress
approved the sum of US$36 million. The government seems to
have money to burn—but, then, it's taxpayers' money, not theirs.

*   *   *

I was trying on a pair of boots in Wayne's Boot Shop in Cody,
Wyoming, when the owner mentioned that predators were causing
more problems now than during any other time in living memory.
As a high school boy he used to shoot coyotes, and he would get
up to US$70 a skin. Now the furrier down the street, just a short
distance from a hotel built by Buffalo Bill Cody and named after
his daughter, Irma, was paying no more than US$10 a skin. The
market had collapsed because there was no longer a strong demand
for fur. As a result, coyotes were becoming more common. "And
there's a lot more bobcat and coons, too," said the boot seller, "and
they're killing off the pheasants and ground-nesting birds." During
recent years grizzlies had also become more plentiful, especially in
the mountainous slab of land between Cody and Yellowstone
National Park. Just the other week one had killed four sheep
belonging to a girl he knew. "The next big problem is just coming,"
he added with a sigh as I was leaving the shop. "And that's the
wolves." He motioned out of the shop. "There's two wolves on the
mountain up there, and two on the other side."

The boot seller was especially unhappy about the way in which
wealthy environmental groups—he mentioned Greenpeace, the
Sierra Club, and the Yellowstone Coalition—were unwilling to
contribute towards the cost of conserving animals such as grizzlies
and wolves, and keeping them away from livestock. Instead, he said,

it was the taxpayer who was footing the bill, and this he believed was unreasonable. I asked if I would find many ranchers around here who would agree with him. "Just stop any rancher walking down this street, and he'll say the same as me," he replied. "Or at least, 90 percent will."

I am sure that he was right. While I was in the West I lost count of the number of times ranchers and wranglers complained about federal expenditure on preserving endangered species, and the failure of the environmental movement to pay toward their conservation. Many argued that this amounted to a subsidy to conservation-minded people, and there is undoubtedly a case to answer.

The Endangered Species Act (ESA) was passed by Congress in 1973, and is primarily administered by the US Fish and Wildlife Service. So far, over 800 species of native animals and plants have been listed as endangered or threatened, and these are subject to a range of protective measures and to programs designed to help their recovery. Mention the ESA to anyone involved with organizations such as People for the USA, and you can see the red mist rising in front of their eyes. In their view, the ESA has been hijacked by what one contributor to *Range* magazine colorfully called "nihilistic wildlife devotees". According to such individuals, it has cost the taxpayer a fortune, damaged the rural economy, and done little to help preserve wildlife.

In many ways, the act is easy to ridicule because it provides protection for some very obscure creatures. Take, for example, the Delhi Sands flower-loving fly, which was discovered in 1984 on a small plot of wasteland in California. The fly was listed as endangered under the ESA, and the medical center that wished to develop the site was forced to pay an additional US$3.3 million to relocate and redesign its facilities in order to protect the flies, of which eight were found. People for the USA worked this out at US$413,744 a fly. Yes, it certainly sounds crazy.

There have been other instances, too, where endangered species protection seems to have been taken to a ludicrous extreme. Take, for example, the Snake River fall Chinook salmon, whose strange history I was told by Dale Kelly of the Alaska Trollers Association, whom I met in Juneau. Washington State's Snake River was once the largest salmon-producing river on the Pacific coast, providing over 600 miles of spawning habitat for Snake River fall Chinook salmon. Then came the dams, which destroyed most of this. During the 1930s, over 70,000 salmon returned to spawn; in 1990, a mere 78 did, and to make it to their spawning grounds they had to get past 16 dams, built, you will recall, with taxpayer money.

Before 1993, Alaskans were harvesting a mere 50 Snake River fall Chinook salmon; yet, under the direction of the Endangered Species Act, Alaskan salmon trollers were forced to give up 11 days of fishing in 1993 to save—and this takes some believing—just one Snake River fall Chinook salmon. That year they gave up 19,000 non-endangered Chinook salmon, just to save that one fish, at great cost to themselves and the local economy.

Critics of the ESA point out that it has led to the collapse of entire industries by locking up land for preservation. Most famously, almost 7 million acres of forest in the Pacific North-West were set aside for the northern spotted owl, and by the government's own admission this led to the loss of 33,000 jobs in the timber industry. The National Association of Home Builders estimates that the cost of a 2000-square-foot home rose by some US$5000 as a result of the restrictions on timber harvesting.

However, there is plenty of evidence to suggest that stories such as this are the exception, rather than the rule. A study by the Massachusetts Institute of Technology (MIT) Project on Environmental Politics and Policy looked at the relationship between the number of species listed under the act in each state and the state's economic performance over a period of 15 years. The study concluded that the data "clearly shows that the endangered species act has had no measurable economic impact on state economic performance." More than 99 percent of the 18,000 projects reviewed during this period under the Endangered Species Act eventually "proceeded unhindered or with marginal additional time and economic costs."

Indeed, the ESA has probably done more good than harm in economic terms. This was an argument that I heard persuasively put by fishing leaders such as Zeke Grader, whom I met in San Francisco. He maintained that were it not for the ESA, the commercial fishing industry in the United States, which provides some 700,000 jobs and generates over US$70 billion a year, would be in a far worse state than it is. Salmon stocks have suffered grievously in the Pacific North-West, not so much as a result of overfishing as the destruction of their spawning sites. The ESA is helping to protect vital spawning habitat and is thus ensuring the survival of species under threat. "Salmon mean business," said a representative of the Pacific Coast Federation of Fishermen's Associations to the House Committee on Resources, "and it pays to protect them. Without the Endangered Species Act to drive recovery, however, you can kiss the entire ·North-West salmon industry—and many other components of the entire nation's fishing industry—goodbye!"

There is no disputing the fact that conserving endangered species is costing serious money. During a single year in the 1990s expenditure on the red cockaded woodpecker was US$5.2 million; on the bald eagle, US$3.5 million; on the grizzly bear, US$5.9 million; and on a small bird called the least Bell's vireo, US$9.2 million. The total expenditure on threatened and endangered species amounted to over US$100 million that year. However, it has been the reintroduction of the wolf that has sparked off the loudest protests in the West.

The gray wolf, as I mentioned earlier, was hunted down and eventually wiped out in the northern Rockies by the mid-1920s. Sizable populations survived in Alaska and Canada, and a small population in Minnesota, which had a hunting season for wolves as recently as the early 1970s. Over the years, many organizations called for the wolf's reintroduction, and eventually, midway through the 1990s, the US Fish and Wildlife Service reintroduced wolves, captured in Canada, to Yellowstone National Park and to locations in northern Montana and Idaho. In biological terms the program has been a remarkable success. Thirty-one wolves translocated into Yellowstone in 1995 bred at such a rate that there were 170 in the area five years later. A similar story can be told for Idaho.

There was, from the outset, plenty of local opposition to the wolf reintroduction program, most of it coming from ranching interests, and scarcely a week goes by when ranching journals fail to highlight the problems that these reintroduced animals are causing. I was especially struck by two stories in *Range* magazine. One told of an Idaho family who were virtually under siege, with a pack of wolves attacking livestock right outside their home. Another recounted how a family camping in the hills of Arizona—there was another wolf reintroduction program in the south—had been attacked by a wolf, which the father shot. "We checked that story out," said an ecologist in the Fish and Wildlife Service's Washington, DC, office when I repeated it to him. "It was complete fabrication. The man had shot the wolf, which wasn't doing anything to his family, and changed his story once he learnt what the penalties were." He added that there was no record of wolves ever attacking humans, although anti-wolf groups—and *Range* magazine—frequently claimed they do.

However, wolves do kill livestock, and it was to assuage ranchers' fears that one conservation group, Defenders of Wildlife, set up a compensation scheme to make good the losses that ranchers suffered. The scheme was established following discussions with ranchers in areas where wolves were beginning to reappear naturally in the northern Rockies. By the year 2000, Defenders had paid out some US$100,000 to over 110 ranchers as compensation for livestock lost to wolves. It now has a trust fund that will continue doing so for the foreseeable future.

Ranchers, and the Farm Bureau, often claim that Defenders fails to honor its obligations and only pays for a fraction of livestock lost to wolves; but Bob Ferris, a large, cheerful character who heads up Defenders' species conservation division, denied this when I saw him in Washington, DC. For one thing, he said, his organization does not verify the kills. Wildlife Services and the Fish and Wildlife Service do. If there is evidence that an animal has been killed by a wolf, they submit the claim to Defenders. Defenders then reimburses the rancher, paying the commercial value for the dead animal. If a spring calf has been killed, Defenders pays what it would be worth at the fall sales. According to Ferris, ranchers tend to exaggerate the number killed by wolves, and he cited a study by Wildlife Services in the northern Rockies which found that only one out of every ten claims submitted by ranchers was a genuine wolf kill. Coyotes, domestic dogs, and other factors accounted for the rest. None of this implies that wolves aren't causing serious problems for some ranchers. In some areas, the authorities have been forced to remove or shoot them. During the time of my visit, the Fish and Wildlife Service was talking of downgrading the wolf from endangered to threatened species. This would give ranchers the right to kill animals that were guilty of killing their livestock.

Ranchers in places such as Cody were too close for comfort to the wolves that were beginning to spill out of Yellowstone National Park. Wally McRae was a good distance further away, in Colestrip, Montana, so the wolves had yet to reach his territory. But he was unimpressed by the bellicose rhetoric which maintained that wolves were going to put ranchers out of business. "The ranchers who are most concerned about the wolf know they've got problems," he said, "and the problem is the markets which they can't do anything about. They need to find a scapegoat, and so it's the environmentalists and the damn wolves. You can go shoot a wolf. You can't go shoot a USDA economist." If people couldn't make it in the ranching business, said McRae, it wasn't because of wolves or coyotes.

Ferris agreed that the wolf was a scapegoat for other ills. "The Farm Bureau likes to focus attention on wolves to take attention from things it cannot do anything about," he suggested, "or from things it does not want to do anything about." The bureau had done little or nothing to challenge vertical integration in the corporate livestock sector, and had even fought against legislation that would have restricted the size of certain operations, and thus benefited family farms. Even a slight decrease in the amount of profit that ranchers made on beef—as little as one US cent a pound—hurt them far more than ten years' worth of wolves ever could, said Ferris. Besides, look at the impact which the forces of nature could have on cattle. During one storm in 1997, over 40,000 livestock perished. It would take many decades for even a huge population of wolves to account for this many livestock.

It seems, then, that the reintroduced wolves are not doing the serious damage to the livestock industry that the Farm Bureau and some ranchers maintain. Certainly, some will suffer, but the scheme established by Defenders is a good example of how conservationists can help to compensate for ranchers' losses. But in fiscal terms, is money spent on wolves and obscure flower-loving flies money well spent? Is it an extravagant subsidy to the conservation elite, as some believe, or legitimate public expenditure?

When I put this question to the wolf expert at the US Fish and Wildlife Services, he laughed and said: "If a rancher who's unhappy about the reintroduction program comes up to me and says, 'This has cost a hell of a lot of taxpayer dollars,' I'll say, 'Yes, it has, but then think of all the taxpayer money spent on gritting the roads and installing US$500 toilets in government buildings, and on many other things.'" Politicians have voted taxpayer money for the program, because they—and, in their view, the public—are in favor of it. The same goes for the appropriations that fund the entire endangered species program. Of course, the politicians have been vigorously lobbied by environmental groups. But this is in no way comparable to the tactics of big business, whose lobbying is backed up by campaign contributions that buy political influence and get the legislation they want.

There appear to be some extraordinary contradictions in government policy. While one department, Wildlife Services, spends millions of dollars a year killing predators, including some 200 wolves, another, the US Fish and Wildlife Service, is spending millions reintroducing the wolf. And when some of the reintroduced wolves do what wolves will always be tempted to do, take a meal at the ranchers' expense, Wildlife Services is often called upon to kill them, and the taxpayer must pay for this, too.

Perhaps we shouldn't be too scornful of such inconsistencies. Governing is often a messy affair, and tradeoffs must be made to satisfy conflicting demands. However, opinion polls suggest that the vast majority of Americans—around three-quarters—are strongly in favor of wolf reintroduction, even though most will never see the wolves, while there is no widespread support for Wildlife Service's lethal control program. The remedy is simple: abolish the latter and persist with the former.

The boot seller in Cody complained that conservation groups were not contributing towards the cost of conserving animals such as grizzlies and wolves. The implication was that they should dig into their own wallets rather than expect the taxpayer to foot all of the bills. As it happens, a growing number of environmental groups are prepared to pay for what they want. For example, the Nature Conservancy in Indiana is giving a group of farmers US$3000 each to help them buy equipment that will help to reduce tillage erosion. In Oregon, a water trust is paying farmers approximately US$1 million to leave water in an important salmon spawning river, to tap other sources of water, and to install better irrigation equipment. And then there is the example of the Defenders of Wildlife, with its livestock compensation schemes for wolves and, in certain areas, for grizzly bears, too. Some animal rights groups see this as the thin end of the wedge. I think this is nonsense: it is a responsible solution to a delicate problem.

CHAPTER 6

FISHED OUT

One afternoon in late January I flew into St. John's, the capital of Newfoundland. A gusty wind laden with coarse crystals of cheek-pricking snow swathed the town in a white blur, and as we slithered our way through the icy streets my taxi driver reminisced about the good old days. When the cod were plentiful, he said, up to 30 foreign vessels would be anchored a short distance up the coast from St. John's, as new crews were flown in from Iceland, Spain, and elsewhere to replace the old. "When the Icelanders were in town," he said with a shake of his head, "they'd tear the place apart. You'd hear stories of them having four hookers at a time apiece! Oh, yes, we loved it. It was great money for us cabbies." Those busy and riotous days have long since gone, as have the cod.

Memories of past times surfaced again when I took breakfast the following morning with an elderly artist who spoke with the lilting accent of Newfoundland, more Irish than Canadian. "At one time you could see ships from all of the nations of the world anchored off St. John's," he said wistfully. "At night, when they were lit up, it was like a city in the ocean. Then the bureaucrats ruined everything." However, as I learnt over the coming days, it was not just the bureaucrats—I think the artist meant the officials within the Canadian Department of Fisheries and Oceans—who were responsible for the collapse of the cod stocks. The fishing industry was culpable, too.

I chose to explore fishing subsidies in Canada, rather than in the United States, for the simple reason that the US government has been relatively parsimonious when it comes to steering public funds towards the fishing industry. In Washington, DC, I spent time with WWF's David Schorr, who has spent several years investigating the scale and impact of fisheries subsidies. Worldwide, they probably amount to some US$20 billion a year. He thought that I would have trouble linking any of the US fishery subsidies to a specific decline in fish stocks. Why not head up to Newfoundland, he suggested. In 1992 cod stocks were so depleted that the Canadian government felt compelled to close the commercial cod fishery, and over 30,000 fishermen and fish plant workers were thrown out of work. Billions of dollars of subsidies had encouraged the expansion of the fishing fleet and an increase in fishing effort. In Newfoundland, I would find a clear link between subsidies and an environmental disaster. I would also find plenty of people with strong views on the subject.

The writer Barry Lopez reckons that fishermen from Bristol, an ancient fishing and slaving port on the west coast of England, were probably fishing for cod off the coast of Newfoundland long before it was officially discovered at the end of the 15th century. "These men, as it were, stepped aside long enough to let the gentlemen discover the land," he wrote, "and then went back to fishing." One of the gentlemen was the explorer John Cabot, who found that the Grand Banks, a great chunk of sea off the coast of Newfoundland, was "swarming with fish." They were so plentiful that they could be caught not only with nets, but with baskets weighted with a stone and lowered into the sea.

The great wealth of fish was a godsend for Catholic Europe, which was suffering from a shortage of protein and needed fish, for religious reasons, on Fridays and during Lent. Soon scores of fishing vessels were sailing across the Atlantic. In 1583 Newfoundland became Britain's first colony, and as the years rolled by more and more Europeans came to make a living from the fruitful seas. By

1770, Newfoundland's European population exceeded 10,000. Over the next century it increased 20-fold, and by 1920 there were over 250,000 Newfoundlanders. Now the population of Canada's tenth province hovers around 540,000. Cod was the catalyst for growth, the raw material that fueled the island's development.

Just over a century ago, Thomas Huxley, one of the greatest Victorian scientists, suggested that whatever we did at sea, it was unlikely to do much harm. "I believe that the cod fishery, the herring fishery, the pilchard fishery, the mackerel fishery, and probably all the great sea-fisheries are inexhaustible," he wrote; "that is to say, nothing we can do seriously affects the number of fish." How wrong he was. Twentieth-century man has plundered the oceans with such recklessness that around two-thirds of the world's most valuable fish species are either overfished, or fished to the limit. There are far too many boats pursuing too few fish, and the recent collapse of the Grand Banks cod fishery illustrates how high technology, poor management, and greed, aided and abetted by government subsidies, can turn the most common of fish into a rarity. At one time well over 1 million metric tons of cod were being harvested each year off the coast of Newfoundland. Today the cod industry is, to all intents and purposes, dead. This is the story of what happened.

The taxi driver had suggested that if I wanted to understand what had happened to the cod, then I could do no better than talk to the Tuckers, a family of fishermen in Quidi Vidi, a small fishing village outside of St. John's. It seemed as good a place as any to start my investigation; so after breakfast I rang Wallace Tucker, who suggested I make my way out to his boat some time that afternoon. I cannot recall ever being colder than I was on the long walk out of town. The thermometer may have been showing minus 10°C, but with the wind whipping off the Atlantic it felt more like minus 50°C. By the time I reached the shelter of Quidi Vidi, whose colorfully painted clapboard houses tumbled down a steep slope to the iced-up water, I was frozen to the core.

The cold didn't appear to bother Wallace Tucker, who was wearing blue overalls and a thin checkered shirt, apparel more suitable for a mild summer's day. A well-built man with a good head of hair and a black mustache, he told me about his life as a fisherman, and about the collapse of the cod stocks. "I was 12 years old, still at school, when I went out fishing with my father for the first time," he explained as he led me into the cabin of his boat. He had now been fishing for 30 years. In his youth there were two fish plants in Quidi Vidi. One had burnt down; the other had been turned into a brewery. There were many more fishermen in those days, too, and there were many more fish. "At one time, before the moratorium," he continued, "I would go out with my father and my brother and we'd take two speedboats out with us as well as the 35-footer as we'd catch so many fish. Cod was what we depended on." During the three years prior to the moratorium, they caught reasonable quantities; but the fish were smaller and younger than they used to be, and each year they arrived in the inshore waters a little later. Something was seriously wrong, and indeed for years the inshore fishermen had been telling the government that there was a problem, that the cod stocks were rapidly declining. Now, belatedly, the government agreed, and in July 1992 the Canadian minister of fisheries declared a moratorium on commercial cod fishing. "We were given a week's warning," recalled Tucker. "So we took our gear in, removed our cod traps from the water, stowed everything away, and wandered along the banks all summer, sucking our thumbs and wondering what to do."

So why did the cod disappear? "You could blame the big draggers from other countries," suggested Tucker, after giving the matter some thought. Some of the draggers were up to 200-feet long—six times the length of the boat that the Tuckers were using in those pre-moratorium days—and one boat could suck up 1 million pounds of fish in four or five days, more than the Tuckers would get in a whole year. The foreign draggers—so called, as they dragged huge nets along the ocean floor—were excluded when Canada extended its fisheries jurisdiction from 12 to 200 miles in 1977. But before long, Canadian draggers took over where the foreigners left off. When the moratorium was declared, politicians and fishery managers suggested that the cod stocks would recover in a matter of a few years. So far they hadn't. Why not? "I'd say the seals are keeping the stocks down," replied Tucker. "They didn't cause the collapse, but there are now too many seals and they must be preventing the cod from recovering. There's no doubt about that in my mind." He pointed out of the cabin window towards a bank covered with snow. "Two months ago a seal was clubbed over there," he said. The man who killed it cut it up for food and sold the pelt. "So far we have killed six this year in the harbor. When we slit one of them open we found four two-pound sea-trout."

Fortunately for the Tuckers, and for several thousand others who used to make a living from cod, they have found a profitable alternative: snow crab, whose numbers seem to have increased dramatically since their main predator, the cod, disappeared. During the early 1980s, crab fetched around 3 Canadian (Cdn) cents a pound, and around half of that was a subsidy given by the government to encourage fishermen to go after crab. Last year, said Tucker, crab was getting up to Cdn$1.92 a pound, and their high value, allied to good catches, meant that the revenue from fishing in Newfoundland was the highest ever, amounting to over Cdn$1 billion. The Tuckers had recently bought a new boat, a 45-footer on which we were now standing, and they could travel far out to sea, much further than they ever used to. For Wallace Tucker, this was a mixed blessing. He was making a good living from crab, but he disliked the new way of life. "In the old days you came back each day," he said. "If you weren't back, then everyone would know to come out and look for you. Now we go out for four or five days and my wife is frightened to death. Everyone is pushing harder and harder. Everyone gets a bit braver, till someone's lost, then you pull back a bit for a while."

When I made my way back toward St. John's, dusk was falling fast, and the descending blackness was pricked by the bright lights of the welcoming city, in whose harbor bars I would drink Guinness and listen to Irish songs, which were now part of these people's folk memory. I got the impression that Tucker, and others like him, were rooted to "place" in a way that is rare in today's fast-moving world. I didn't ask how long his family had lived here, and fished these waters, but it may well have been for many generations. Later, when I visited a fishing village further along the coast, a fisherman asked me where I came from. England, I said. "My family too, we're from England, from Dorset," he replied. I asked when they came. "Oh, I'd say about 350 years ago," he replied cheerfully. They had been here ever since, living off cod and the bounty of the sea.

Tucker had said nothing about subsidies, or the role they played in the decline of the cod. Not that it mattered. His reticence on the subject was more than compensated for by Professor Bill Schrank, who taught at Memorial University, a sprawling campus of unprepossessing modern buildings overlooking St. John's. Schrank was an economist and a native of New York, and in terms his appearance and outlook, his manner of speaking, and his bookish air, he was light years away from the life at sea. As soon as I arrived at his office, he presented me with a paper he and an economist from Louisiana State University had written entitled "The Concept of Subsidies", together with two papers he had written on the Newfoundland fishery and the cod crisis. It was, he said, an immensely complicated issue. A whole range of factors could have led to the collapse of cod stocks; but there was no denying that subsidies, and a determination by government to keep as many people fishing as possible, lay behind the crisis.

To understand what went wrong, Schrank said that I needed to understand the importance of cod fishing in the island's history. Here, briefly, is what he told me.

By the 1880s, virtually all of the available coves along Newfoundland's craggy coastline had been settled by families whose living came from the sea. There was little in the way of formal education and sons joined their fathers in the boats as soon as they could. Indeed, as recently as 1991 almost one quarter of the population was illiterate, and the lack of educational achievement meant that many families never looked for a living beyond the fishing industry. In 1946, Newfoundlanders voted to join Canada, persuaded, in part, by the fact that they would benefit from family and other welfare allowances paid for by the federal government in Ottawa— allowances that the provincial government could never afford. In 1957, unemployment insurance was extended to fishermen. To qualify, fishermen and fish plant workers needed to get between 10 and 12 stamps a year, each stamp representing one week's work. Once they had these they could claim unemployment insurance for the remainder of the year. These payments, together with a whole gamut of subsidies, helped to create the present crisis, according to Schrank.

In fact, over the past 30 years there had been four major crises, and Schrank proceeded to spell them out, one by one. The first of these he attributed to the remarkable advance in fishing technology after World War II: bigger and better nets, larger vessels and the introduction of sophisticated fish-finding devices enabled fishing fleets to catch ever greater quantities of fish. Cod catches off Newfoundland, largely by foreign distant water fleets, rose from 700,000 metric tons in 1950 to almost 2 million by 1968. Stocks and catches swiftly fell thereafter. In response, over the next five years, the number of registered fishermen fell in Newfoundland by one quarter. The Canadian Department of Fisheries and Oceans (DFO) recognized that there were too many fishermen and wanted to reduce the number by half. However, other government departments, bending to pressure from the fishing communities, refused to cooperate in reducing the numbers involved in the industry.

The second crisis was sparked off by the rise of oil prices in 1974. This led to a worldwide recession and decline in the price of fish. Many of Newfoundland's fish processing firms came close to bankruptcy, and the inshore fishing fleet was rescued by government subsidies. There was another crisis in 1981, when recession in the United States again led to a collapse in prices. This time, the government saved the industry from oblivion by nationalizing many of the province's fish processing firms. The fourth crisis, in Schrank's scheme of things, was the present one, brought about by the collapse in cod stocks.

"If the government had acted sensibly in 1977, when it extended jurisdiction over fisheries from 12 to 200 miles," reflected Schrank, "Newfoundland could have had a viable fishery." But instead of allowing the cod stocks, hammered by years of overfishing, to recover, federal and provincial governments sprayed subsidies around as if there was no tomorrow. Partly as a result, the number of registered fishermen more than doubled within four years, as did the number of inshore vessels. Meanwhile, store-based freezer capacity more than doubled, too, and the amount of unemployment insurance paid to Newfoundland's fishermen and fish workers rose by a factor of four. The fishery had become, in the parlance of those opposed to subsidies, a stamp fishery. Schrank reckoned that during the period of 1972/1973 to 1990/1991, the federal and provincial governments spent nearly Cdn$4 billion on the fishery, and over half of this went on unemployment insurance payments.

During the early 1980s, Schrank and fellow economists at the university constructed an economic model that sought to match fish stocks, managed sustainably, to fishing effort. They concluded that the fishery could not support more than 6000 jobs. Schrank said he now thought this to be an overestimate, and that in future the fishing industry might support no more than 3000 people. Distilled into a couple of lines, Schrank's argument was this: subsidies, unemployment benefits, and emergency bail-outs for fishermen, fish plants, and banks in difficulty, led to overcapacity in both people and equipment, and, as a result, to overfishing. A commercially viable fishery is one which does not depend upon a continuous drip of outside assistance, and where the numbers employed make a reasonable living without threatening fish stocks. It was time for the government to bite the bullet, and use its financial clout to ease large numbers out of the fishery, rather than retain far too many in it.

If Schrank had to apportion blame to the collapse of cod stocks, where would it lie? There was a long list of possible causes, he said. Most obviously, there was overfishing by foreign fleets, and over-fishing by Canadians once the foreigners left. Some scientists had suggested that the sea was much cooler, and this could have affected stocks, and the science on which quotas were based had apparently been faulty. Seals were blamed by some, and so was the lack of capelin on which the cod feed. The pressure put on government by fishermen and fish corporations to allow higher quotas than made biological sense proved irresistible at times, and the weakness and indecisiveness of the government had not helped. As a result, a whole combination of factors may have been involved. "I see it as a Greek tragedy," concluded Schrank. "No one is evil, but everyone is acting in their own best interests, and without rigorous manage-ment the whole thing moves inexorably towards the destruction of the fishery."

Reg Anstey, the portly, quietly spoken secretary of the Fish, Food and Allied Workers Union, agreed that various factors had led to the decline of the cod. "Who was to blame?" he reflected in his office overlooking the harbor. "Well, I think there's enough blame to go round for everybody. Bad practice in the inshore fleet. The offshore draggers, both foreigners and our own. Then the drop in water temperature affected the cod. And in the days before the moratorium the government tended to throw fishing licenses around to sort out unemployment." But Anstey was optimistic about the future now. The government had bought out the licenses of 1500 boats, so the resources were being shared by fewer people, and the snow crab fishery was proving a great boon to many fishermen. Since the moratorium the number of fishermen had been reduced to 14,000, and the fishery was now much better managed than it was in the past.

When I told Anstey that Schrank was adamant that the fishery could only maintain between 3000 and 6000 fishermen, he snorted in disgust. "That's a ludicrous opinion," he said. "He couldn't be further out to lunch. Anyway, I've never heard of this fellow, and there's probably a good reason why!" But what about all of the subsidies that had encouraged overcapacity? "There are no subsidies going into the fishery now," he said emphatically. But what about the unemployment insurance? "That's not a subsidy," he replied curtly. "The funds for that are paid for by employers and employees, not by the taxpayer."

It came as no surprise that Anstey's and Schrank's views of the fishery's future were so at variance with one another. Anstey and his colleagues at the union had a mandate to represent their members, the vast majority of whom wished to remain in the industry. You could argue that they had a vested interest in retaining high numbers within the fishery, even if it meant doing so at the taxpayers' expense. Schrank, on the other hand, did not have to worry about the sensibilities of those whose livelihoods would be lost if the government chose to downsize the industry, although he was not arguing for an immediate cessation of all state aid. Rather, he wanted the government to provide generous enough payments to help the industry move from being bloated to sleek, without causing major hardship to those individuals who were pushed out of commercial fishing.

The provincial fisheries minister, John Efford, was in absolutely no doubt about what was to blame for the collapse of the cod stocks. "I'll give you one word for it," he said dramatically. "Greed. It was greed that did it. The greed of the foreign, Canadian, and Newfoundland factory freezer trawlers was entirely to blame for the collapse of the cod."

Mention Efford, a balding, bespectacled character whose accent becomes progressively more Irish the more excited he becomes, to environmentalists and they bridle with anger. Here is the man who famously stood up in the House of Assembly and declared that he would like to see the "six million seals or whatever number is out there, killed and sold, or destroyed or burned." But Efford is much admired by taxi drivers, many fishermen, and the man with a pint of Guinness in his hand: likable, intemperate, and outspoken, he might have stepped straight out of a novel by Flann O'Brien.

"Pressure from the big companies led to an increase in quotas, far beyond what the stocks could sustain," explained Efford. "There was huge pressure and lobbying from the big companies, and the politicians continually ignored the advice of scientists and the inshore fishermen who said their catch was going down because of the activities of the offshore fleet." During one year, scientists at DFO had recommended a cod quota of 275,000 metric tons, but the federal minister in Ottawa opted for 400,000 metric tons. The collapse of the fish stocks, in Efford's opinion, had nothing to do with either the smaller inshore fishermen, operating from boats of 35 feet or less, or changes in temperature, or even the seals, though like most fishermen I met, he was convinced that seals were hindering recovery of the cod stocks.

The biggest of the big companies that Efford referred to, and the largest fish processing plant, was Fisheries Products International Limited (FPI), which rose, phoenix-like, from the ashes of the economic crisis of 1981. With the dramatic fall in cod prices, and the loss of US markets, many heavily overborrowed fish processing companies found themselves in dire financial straits. The Canadian government, mindful of the jobs that could be lost, put together a rescue plan, and with Cdn$250 million of taxpayers' money four bankrupt deep-sea fish companies from Newfoundland and Nova Scotia were reconstituted as FPI. At one time, FPI had over 50 trawlers operating off the coast of Newfoundland, but since the 1992 moratorium it had sold most of these. However, it remained in rude health, largely because it had managed to source cod and other fish from elsewhere. In 1992 it had sales of Cdn$250 million; by 1990 these had risen to Cdn$700 million.

As the company had been a major player in the cod fishery, I was keen to talk to its chief executive, Vic Young. My pleas for an interview fell on deaf ears, which came as no surprise to Owen Myers, a lawyer who worked in his younger days as a fisheries observer on both foreign and Canadian draggers operating off the Grand Banks. "Of course Mr. Young won't see you," he said as we sat down to dinner, a moose stew which he and his girlfriend had cooked. "Mr. Young was awarded the Order of Canada—that's our highest honor—for destroying fish stocks!" Myers had an acerbic wit and was outspoken in his criticism of the politicians and businessmen whom he believed had helped to wreck the Newfoundland cod fishery.

In Myers's view, the first major blow to the cod came during the 1960s, when foreign fleets with German technology began to exploit the cod stocks under the ice. This was a method of exploitation he witnessed with growing horror when he became a government fisheries observer in the late 1970s. "For weeks on end," he recalled, "all you'd hear was dong-dong-dong as the reinforced hulls crashed through the ice field." The draggers were huge boats, 160 feet plus, and they would catch huge quantities of fish. Prior to their arrival on the scene, the ice sheets protected the cod in the winter, but the draggers could now fish 365 days a year, and for the cod there was nowhere to hide. "By 1977 all the big fish had been caught by the foreign draggers," continued Myers, "and there were only the young left. That's what our draggers were catching—cod aged four to five years old that should have been left alone to spawn." Myers said he remembered an order that came straight from the head office of FPI: "'Don't bring back small fish,' it said. The only thing that mattered to them was to keep the processing plants at maximum capacity, to make as much profit as possible. If it meant throwing smaller fish overboard, that made no odds to them. It was an environmental *Apocalypse Now.*" The practice of throwing back smaller fish, and keeping the larger—it is known as high-grading— was witnessed by Myers and many other observers. "Sometimes they would let a whole bag of smaller fish go back to the sea, and as often as not they would all have been killed in the net." Of course, these discarded fish were not counted as part of the quota.

"But if you were an observer," I asked, "why didn't you do something about it?"

"I did!" exclaimed Myers indignantly. "I was reporting what was happening all of the time, but the government took no notice. The whole system was totally corrupt from top to bottom." Myers was robustly critical of the union, which he described as the crowd control division of the Ministry of Fisheries. "The union was built around the dragger crews," he said. "How can you have a body which represents both the draggers, who were wiping out fish stocks, and the small inshore guys whose livelihood was being threatened by the draggers? You can't. It's a nonsense." He noted that the same small coterie of individuals had run the union for almost 30 years, and he was scathing about the cozy relationship that existed between the union, the bureaucrats who were supposed to manage the fisheries, and big business. "The man who used to be in charge of FPI's dragger operations is now in charge of the fisheries conservation board," he said. "That's like putting Heinrich Himler in charge of the resettlement of the Jews." He said that he thought that the small guys—the inshore fishermen—were probably no better in terms of their attitude to fish stocks than the big guys. "But at least they don't have the capacity to behave so badly," he added.

*   *   *

In virtually every story I had so far investigated, subsidies had benefited the rich more than the poor, the big industries more than the small, the politically powerful over the politically weak. But who were the villains here? Were they really just the big companies, and the owners of the draggers, or had the numerous small fishermen, many dependent upon welfare handouts for much of the year, pushed the cod towards extinction, too? To find out what they, the small operators, thought, I headed down the magnificent, heaving coastline to meet Tom Best, a member of a small fisheries cooperative. He began by telling me about the history of the port.

When Best was growing up in the picturesque fishing port of Petty Harbour, a short distance along the coast from St. John's, there was a clear division between the descendants of Irish immigrants, who lived on one side of the narrow inlet where the boats were moored, and the descendants of the English, who lived on the other. "We used to think twice about crossing the bridge," recalled Best, a sturdy, ruddy-faced man in his fifties. The enmities were of long standing. Petty Harbour had first been settled during the early 1500s, in the days before Newfoundland became a British colony. "There used to be two hockey teams," continued Best, "one for the Irish Catholics, another for the English Protestants, and they both lost all their matches. But my generation put an end to all that crap." Now there was one hockey team, and a winning one at that, and this new-found cooperation within the community had extended to the fishing community, too. Petty Harbour was home to the few non-unionized fishermen in the province, and the local fisheries cooperative, established in 1984, had a core of 80 committed members. Best maintained that the structure of the union had meant that it had failed to take a strong stance against the draggers who were responsible for overfishing the cod. "We started to notice there were problems as long ago as the 1960s," recalled Best. "Then the foreign fleets were excluded and our own Canadian draggers continued to damage the fish stocks. If we inshore fishermen had been here for another two million years, we could never have had the impact those people had. They destroyed in 30 years what we had not in 400."

I mentioned to Best that among those individuals I had seen in St. John's was Dr. Art May, who had spent some 20 years working as a scientist in DFO before becoming the deputy minister—in other words, the chief bureaucrat—in the federal Ministry of

Fisheries during the early 1980s. May conceded that the draggers had had a serious impact on fish stocks, but he also believed that the inshore fishermen were to blame, too. He felt that the unemployment insurance had created "a bonanza of unprecedented proportions," and encouraged far too many small inshore fishermen to remain in the industry. The people of Canada were paying hefty sums to keep tiny communities in Newfoundland alive, he had told me. "It's like keeping the horse and carriage industry going in the face of the automobile," he said. Like Schrank, he believed that there were far too many small fishermen: it was time to ease them out.

Best listened politely, but grim faced, while I transmitted the guts of my conversation with Dr. May. "He was one of the most destructive persons in the fishing industry," he replied with barely concealed anger. "He was the deputy minister who supported all these big companies who were doing the offshore fishing." May's attitude still permeated government thinking, according to Best. "Their attitude is that the problem has nothing to do with technology. It's all because there are too many people involved." Just look at the buy-out programs, said Best. Nine-tenths of those bought out so far had been small fishermen; the larger boats, the ones with the greatest ability to catch fish, had hardly been touched by the program.

But surely, I suggested, there was something seriously out of kilter here. Billions of dollars had been paid to fishermen in unemployment insurance, and it was financial madness to reward large numbers of people with handouts in perpetuity, providing they worked for a couple of months a year?

"Thirty years ago," replied Best, "we had a fishery that operated maybe six or seven months a year, generating money for ourselves and the province. Now it lasts three or four weeks. Why? Because government policy destroyed the resource." By mismanaging fish stocks, and by subsidizing the industrial fishing fleet, the government had brought about the collapse of the cod fisheries. It was to blame for the fact that the inshore fishermen were now claiming unemployment insurance for such lengthy periods. Best was stunned by what he saw as an implied criticism: that fishermen were playing the system. As a young boy he had learnt another trade—he was also an electrician—to ensure that he could find work when there was no fishing. Prior to the cutback in seal quotas —largely as a result of pressure from European and US-based animal rights organizations—he and many others in Petty Harbour, and elsewhere along the coast of Newfoundland, had joined the annual seal fishery for six or seven weeks each year once the cod quota had been caught.

Best was understandably worried about the future for small fishermen. Ideally, he wanted to see a diverse fishery, with cod, lump-fish, crab, lobster, and seals providing many months of employment, rather than just a few weeks. He was particularly concerned by the latest trend, which saw small inshore fishermen upgrading to larger boats of the sort that the Tuckers now had, the 45-footers and upwards, which had the capacity and technology to catch large quantities of fish very quickly. The union, said Best, was complicit in the emergence of this new midshore fleet, and was doing nothing to help the inshore fishermen. Above all, Best wanted to see better management of the fish stocks, and he felt that the Petty Harbour fishermen's cooperative, of which he was a former president, had shown what could be done, albeit on a small scale. "This is the only community in Canada that has banned gill nets," he said with pride. "Every year tens of thousands of gill nets are discarded or lost. It's a very destructive technology." The ghost nets, as these discarded nets are called, continue to catch fish, dolphins, and other sea life, and the trapped animals die and rot. Of course, he added, theirs was a relatively modest achievement because it only applied to the waters immediately around Petty Harbour; but it was indicative of the desire among many inshore fishermen to conserve fish stocks. However, there was little they could do themselves to improve the overall management of the fisheries off Newfoundland.

According to Best and to many other inshore fishermen, the government had persistently ignored warnings about the state of the cod stocks. "Yes, some inshore fishermen were complaining," agreed Art May. "But how many? Five percent?" In May's view, the vast majority of inshore fishermen did not complain about declines in stocks. Ray Andrews, a senior civil servant at the time of the cod collapse, disagreed. "We did not listen nearly enough to the inshore fishermen," he said when I visited him at his home in St. John's. "The offshore fishermen—and the scientists who were basing their findings on the offshore catches—were telling us that there were plenty of fish still, and that's what we based the quotas on." In fact, the cod were declining, but the decline was not reflected in the catch. The draggers, with their fish-finding devices and advanced technology, were simply getting better and better at finding the fish. Dr. George Winters, one of the chief scientists at the DFO, agreed that the scientists depended far too much upon the offshore catch data, and he was among the first to realize that quotas had been set far too high. Provincial fisheries minister, John Efford, was also adamant that the government had failed to heed the warnings of the inshore fishermen: "They were crying out that something was wrong," he said. "They were right, and their advice was ignored."

Responsibility for the management of the fishing industry in Canada is split. While the federal government is in charge of affairs at sea—at setting quotas and the like—the provincial government has responsibility for land-based matters and, in particular, for regulating fish processing plants. During the years when cod stocks were in decline, both were providing subsidies. The federal government, for example, provided cheap loans to build boats and improve gear, while the provincial government poured money into the financing of processing plants.

Efford was in no doubt that all subsidies were, as he put it, stupid. "I don't believe in subsidies," he said bluntly, "and since 1996 not one nickel has gone into any fishery in Newfoundland. If it is profitable and viable, you shouldn't have to finance it." Efford, needless to say, did not consider unemployment insurance to be a subsidy. He had no doubt, however, that subsidies helped to cause the crash in fish stocks. "Subsidies in the 1970s and 1980s led to the building of far too many fish plants," he said. "By building too many it created an appetite for fish. When I took over in 1996, there were 245 fish plants. Now I've got the number down to 117."

However, in Efford's view, the federal government had been responsible for handing out the bulk of subsidies, and for failing to put the brakes on industrial fishing. This was a view that was vigorously contested by Art May, who said he found this analysis "shallow and convenient". In fact, May said, the federal government had continually battled against the province, which was constantly lobbying for greater expansion. Back in 1977, according to May, he had persistently stated that there were enough fishing boats to catch all the available fish; but while his department, the DFO, preached restraint, others—including the Department for Regional Economic Expansion—were doing the opposite. "It was the provincial government, not the federal government, which was responsible for licencing the fish plants," he said.

Whoever you chose to believe, neither the federal government, nor the provincial, emerges from the wreckage of the cod industry smelling of roses.

*   *   *

The story of Newfoundland's cod crisis has already become a classic of sorts. It has provided a parable about the tragedy of the commons, exemplifying our failure to manage communally owned resources in a sane and sustainable way. Sadly, it is one example among many, and had I the time, I could fill several volumes with similar tales of oceanic plunder. Here, plucked almost at random from my experiences, are a few other salutary reminders of what we have been doing over the past few decades.

During the early 1970s, I decided to look for a summer job in the English port of North Shields. I made inquiries in the offices of trawler companies along the busy quayside, and was told that there were jobs to be had, but that I had to expect at least six weeks away in the cod-rich waters around Iceland. This was in the days before the third Icelandic cod war, and Iceland's subsequent decision to declare sovereign rights over waters within 200 miles of its coast. I declined the offers, once I realized how tough a job it would be, and instead found work on a Yorkshire farm.

In those days North Shields harbor was heaving with life, and scores of trawlers jostled for space. Today the only job you would find on the quayside would be selling ice creams or hot dogs. The fishing industry has all but disappeared, and the North Sea coast is dotted with fishing ports where there are now few commercial vessels left afloat. Several factors led to the collapse of the fisheries. The European Union's (EU's) Common Fisheries Policy, arguably one of the most disgraceful, expensive, and environmentally destructive management regimes ever conceived, was partly to blame. UK waters were opened up to other fleets, and at the same time national and European subsidies encouraged the rapid growth in fishing capacity. Before long, too many boats were competing for too few fish, and ultimately this led to the collapse of many fish stocks. Here, as elsewhere, the smaller, less capitalized enterprises went to the wall first.

In the Philippines I spent time interviewing peasant fishermen who fished with hook and line from small outrigger canoes. Most lived in flimsy shacks with no running water and insufficient food to feed their families, not least because they caught a fraction of what they, and their forebears, used to catch. The decline in fish populations had little to do with them. It was primarily a result of a deal, made by their corrupt president, whereby Taiwanese trawlers gained access to Filipino waters in return for large sums of cash— which went, needless to say, to President Marcos and his cronies, not to the tens of thousands of fishermen whose livelihoods were being ruined by foreign overfishing.

In Kerala, a verdant state in southern India whose shores are lined with palm groves and fishing villages, I saw poverty every bit as serious as you would find in the urban slums of Delhi or Mumbai. Prior to the 1960s, there were no mechanized boats in Kerala's waters, although the seeds of change had already been sown by a foreign aid project that helped to establish a lucrative export trade in tiger prawns. Before long there was a scramble for "pink gold" and a new class of absentee trawler-owner appeared on the scene. In 1971, 90 percent of the catch was taken by peasant fishermen, fishing from canoes. By 1980, the total catch had almost halved, as a result of overexploitation, and 100,000 peasant fishermen saw their share plunge, while the owners of 3500 trawlers grabbed the bulk of the catch. When I first visited Kerala there were frequent outbursts of violence, with the peasant fishermen setting fire to trawlers, whose owners exacted their own forms of heavy-handed revenge. In order to compete with the trawlers, many traditional fishermen have fitted outboard motors to their canoes, and over the years these have become bigger and more powerful. Since few fishermen had the means to buy the motors, they were forced to borrow from money-lenders, often at exorbitant rates of interest. As a result, many fishing families are now deep in debt. In the meantime, fish catches have continued to decline. If there is a moral to this story it is this. When the rich and powerful, and those with the most voracious fish-catching equipment, are left to do as they please, fish stocks will almost always suffer, and so will those who make a living by using "less efficient" means.

During the late 1980s and early 1990s, environmental groups took relatively little interest in the fishing industry. They were more concerned with whales and dolphins, seabirds and sea otters—in short, with the charismatic species that attract members and public interest. But times have changed, and WWF—formerly known as the Worldwide Fund for Nature—has taken the lead in alerting the public and governments to the unfolding disaster at sea. One of its most articulate spokesmen is David Schorr, who runs WWF's endangered seas campaign subsidies initiative. "The root of over-fishing is not subsidies," he said when I met him in Washington, DC. "It is too much uncontrolled fishing. Technology is a key factor, as is the failure to control fishing effort. The basic problem is inadequate fisheries management, and subsidies are just part of the story."

No one can dispute that there are now too many boats pursuing too few fish, and many governments recognize this. In early 1999, delegates from almost every significant fishing nation agreed on a plan to reduce fishing capacity, starting within six years. According to WWF, the global fishing fleet may be two and a half times the capacity required to fish the oceans at a sustainable level. In just one 20-year period, from 1970 to 1990, the world's commercial fishing fleet doubled in size. At the same time, boats became bigger and faster, devices to locate fish became more sophisticated, and synthetic nets became larger and stronger. According to a recent analysis of the fishing industry by the United Nation's Food and Agriculture Organization (FAO), the efficiency of fishing vessels has increased by 1 percent to 3 percent a year over the last 30 years. "There are no reasons to believe that this trend will not continue," says the FAO ominously.

A century ago, the world's marine fish catch was some 3 million metric tons a year. By 1989 it had risen to 85 million metric tons, and during the 1990s it fluctuated between 80 and 85 million metric tons. Had our fisheries been managed sustainably, according to the FAO, we would have had the chance of catching 100 million tons a year. Half of the major fish stocks are now fully exploited, 15 percent are overfished, and 7 percent are depleted. Although the catch remained stable throughout the 1990s, its composition changed, with the catch of high-value bottom-living and large pelagic species falling as a result of overexploitation, and the catch of shorter-lived, surface-dwelling schooling fish rising.

Industrial fisheries—that is to say, fisheries that are designed to yield fish oil and animal feed—now account for around one third of the annual harvest. These fisheries have led to the decline of anchovies off Peru, pilchards around Japan, capelin in the Baring Sea, sardines off California, and sand eels in the North Sea. These species are the building blocks of the food chain, and with their decline we have seen the decline in species that grace the table. Even more scandalous has been the practice of discarding unwanted fish. To the 85 million tons of landed catch a year, we should add another 18-40 million tons of "bicatch". This is a euphemism for the unwanted fish—wrong species, wrong size, wrong sex, wrong time—that are thrown back in the sea once caught, mostly dead.

Subsidies may not be the root cause of overfishing, as Schorr pointed out, but they are very much part of the story. Over the past few years, WWF has helped to push them onto the political agenda, along with the World Bank and several nations, who for economic as well as environmental reasons consider them to be undesirable. Establishing the scale of fishery subsidies is far from easy for a number of reasons. Governments themselves are often unaware of their size and nature. "In the United States," said Schorr, "the federal govern-ment often has no real idea how its money is being spent at the local level, and the data is very dispersed and hard to find." And in countries such as Spain, many of the subsidies are regional subsidies, and the national government is not always involved in dishing them out. Some subsidies—those, for example, that go towards fleet renewal, export promotion, port facilities and gaining access to foreign waters—are budgeted. Others—such as deferral of income taxes for fishermen and exemptions from fuel tax—are not.

So far, the most thorough study of fishery subsidies has been carried out by Matteo Milazzo, a US fishery economist, and it is worth briefly summarizing his findings. In total, suggests Milazzo, domestic budgeted subsidies that are effort and capacity enhancing, and which consequently often lead to overfishing, may be as high as US$3.5 billion globally. Among the major players are the European Union, Japan, and China, with the United States some distance behind in the profligacy stakes.

To this we must add a further US$1 billion in budgeted subsidies that are designed to gain access to foreign waters. For example, in 1998 US taxpayers paid US$14 million to enable 40-odd US tuna vessels to fish in the South Pacific. This is what the US government paid the Pacific island states; in effect, this was a subsidy worth around US$300,000 to each tuna boat. More reprehensible, and far more significant, are EU subsidies to gain access to foreign waters. Over the past decade the EU sought to reduce overcapacity in the commercial fishing sector—overcapacity, let us not forget, that was encouraged by subsidies—by exporting its ability to catch fish. In 1996, for example, the EU signed a five-year agreement with Mauritania, a vast, sparsely populated chunk of the Sahara Desert in West Africa. In return for US$350 million, the Mauritanians have allowed over 240 mainly Spanish vessels to catch 182,000 tons of fish a year in their fish-rich waters. A similar deal was made between the EU and Argentina, with the EU paying US$162 million to allow European boats to fly the Argentinian flag and exploit the local hake stocks. In both cases, the subsidies have had a serious impact on local small-scale fishing enterprises and on fish stocks. In Argentina, the hake fishery collapsed as a result of overfishing. Off the coast of Mauritania, several fish stocks have gone into decline.

To these budgeted subsidies we must then add the various unbudgeted subsidies, and those that Milazzo describes as cross-sectoral: aid to ship-building, shoreline preservation, subsidies for infrastructure, and the like. All of this benefits other sectors of the economy, too. Altogether, subsidies add up to around US$21.5 billion a year, or 20–25 percent of world fishery revenue.

Milazzo divides subsidies into two types: the bad and the good. The former enhance capacity and fishing effort, and lead to overfishing. This is precisely what happened in Newfoundland. The latter help to sustain and conserve fish stocks. The bad vastly exceed the good. Subsidies that have a positive environmental impact amount to no more than 5 percent of all fishery subsidies, and even those whose intention is to limit fishery effort—for example, by decommissioning boats—have often failed to work. According to Schorr, the bulk of subsidies have tended to go to the largest companies, and thus to those sectors within the industry that have the greatest capacity to catch fish and to overfish.

Unfortunately for WWF, the fracas in Seattle, when rioting disrupted the World Trade Organization (WTO) meeting in the fall of 1999, drew attention away from an important subsidy-tackling initiative that it was promoting. "We had the highest-level press conference on an environmental issue," recalled Schorr, "and it was devoted to tackling fishery subsidies." Attending the conference were foreign ministers, senior US government officials, and diplomats from several nations who were determined to get fishery subsidies onto the upcoming WTO agenda for the talks. Among those nations who were calling for the WTO to agree on rules that would reduce and eliminate subsidies that contribute to overfishing were the United States, Iceland, Norway, New Zealand, Australia, the Philippines, Peru, and Argentina. Iceland, New Zealand and Norway have already reduced their subsidies considerably, and it is time others followed suit. "What we need," said Schorr, "is a binding text similar to the one that we have for agriculture, which will eliminate certain types of subsidy, and particularly those that encourage or lead to overfishing. We're also fighting hard to increase transparency in the WTO, and to get governments to tell the truth about the levels of subsidy."

In the overall scheme of things, fishery subsidies are dwarfed by subsidies to the automobile, to energy producers, and to agriculture; but there is no getting away from the fact that they have had a malign influence. By favoring the big players, they have discriminated against the small. The industrial fleets have always won out at the expense of the far more numerous smaller boat-owners. Abolishing all capacity-enhancing subsidies will not be enough to solve the problem of overfishing, but it will certainly help.

In the waters off Newfoundland it was the big draggers, working throughout the year and catching vast quantities of fish, which were primarily responsible for the collapse of the cod. As Tom Best said, the inshore fishermen could never in millions of years have done what the draggers did in a matter of a few decades. They simply don't have the technology to be so destructive, although the emerging mid-shore fleet is proving highly efficient—efficient in the sense that large numbers of fish can be quickly caught with relatively little manpower. But is that the only sort of efficiency?

"You can't stop technological progress," Art May told me. "It's an inevitable law of nature." But what if the nations, or individuals, whose task it is to manage fish stocks, and to ensure that they provide a sustainable harvest in the future, fail to adequately control the latest generation of high-tech trawlers? This is precisely what has been happening in many parts of the world. Although there is no way of preventing net makers and boat-builders and sonar manufacturers from designing more efficient products, the time may come when we should insist that certain types of vessel must not be allowed to fish in certain waters. "To have the society we want," suggested David Schorr, "we may need to willfully embrace inefficient technologies. After all, what is wrong with that?" Nothing, though I might quibble with the use of the word inefficient. Is a fishery that provides a modest living for 10,000 people working from small boats using hook and line less or more efficient than one of similar size that provides a living to 100 individuals using large trawlers with the most modern gadgetry? I would say it is more efficient: the same resource provides a living for a greater number of people. I am not suggesting that all fishing should be conducted from canoes. If that were the case, fish stocks far offshore, would remain unexploited. However, we do need to rethink our ideas about efficiency —and get rid of the perverse subsidies that encourage overfishing. If we don't, then there will be fewer and fewer fishermen making a living from the sea, and fewer and fewer fish.

7

HIGHWAY ROBBERY

I headed across the Appalachian Mountains one cold spring day, and arrived in Elkins, West Virginia, late in the afternoon. There was still enough light in the sky to see that this was a handsome little town. An historic center with a grid of streets lined with old wooden buildings lay between a rushing river and the grounds of a venerable old college. Beyond lay the suburbs, and the inevitable jumble of supermarkets, fast food restaurants, and car parks. Surrounding all this was fine countryside that was struggling out of its winter apparel. Snow still hung to the high peaks, and the poplars and cherries along the riverbanks were only just breaking into bud. It was still far too early for the tourists who would flock to Elkins when the warmer weather came, and the presence of a stranger in Beander's Bar immediately attracted the attention of the huge blond-haired barman. What was I doing in Elkins at this time of year, he inquired as he pushed a pint of Guinness across the bar?

I had come to find out about the proposed Corridor H highway on the advice of David Hirsch, transport campaigner at Friends of the Earth (FoE) and one of the co-authors of *Road to Ruin*, which lists "the 50 worst road projects in America that would waste tax dollars, harm our communities and damage the environment." Before I left Washington, DC, Hirsch ran me through the projects that he considered to be the least justifiable, and at the top of his list—"pure pork", as he called it—was Corridor H, a proposed 100-mile, four-lane highway that would slash through the Appalachians of West Virginia and cost over US$1 billion.

I mentioned to the barman that among those whom I had arranged to see here were members of Corridor H Alternatives, an organization that had successfully helped to stall the building of the highway.

"God-damned bunny-huggers!" he exclaimed. "You know, there's times when one of those environmentalists sits right where you are and I just feel like leaning over and. . ." He leant across the bar, made as if to put his shovel-sized hands round my neck, and went through the motions of squeezing and shaking. Then he laughed, and held out his hand. "I'm Beander," he said. "This is my bar." His real name was Daniel Kwasniewski. "Yes," he continued, "I'm the second cousin of the president of Poland. My grandfather came here to the United States and his brother, the president's grandfather, stayed in Poland. Just imagine if it had been the other way round!"

During the course of the evening, Beander, an endearing character despite his monumental build, and many others who drank in his bar, or played with the salacious machines that required punters to match pairs of naked breasts with faces, told me in no uncertain terms why Corridor H was needed. They said that the vast majority of people who lived in Elkins wanted to see it built, and that I would soon discover that the entire political establishment of West Virginia was in favor of the road. The bunny-huggers, or salamander-worshippers as I later heard a pro-road campaigner call them, were mostly outsiders who had come to live in the area, and they cared little about the poverty and joblessness that were a feature of the state.

As I was leaving, Beander suggested that I head down the street to a bar called Jabberwock. "That's where the hippies hang out," he said. "They'll give you the other side of the story." In fact, Jabberwock was closed, so I returned to my bed-and-breakfast. Its owner, Harry Henderson, was a regular at Jabberwock, and although he had a pony tail you would hardly describe him as a hippie. A retired US naval captain, he now taught business studies at the local college. He said it was possible to have a rational conversation with the college students about Corridor H, but not with the adults outside. "There's just no common ground," he said, confirming the impressions that I already had, having looked at the various web sites that champion or oppose Corridor H. One of these was dedicated to the views of the militantly pro-road Corridor H Action Committee. It was clear that in the minds of its supporters the building of the new road was inextricably bound up with the survival of democracy. This was championed in almost mythic terms: "We are the common people who have come together in an effort to show that democracy in America exists and it remains vibrant. . . . We are a walking civics lesson. . . . We are showing our reverence for the dead, to those who have sacrificed, sometimes sacrificed their very lives, for the cause of American democracy." Emotions were clearly running high.

Visit the offices of virtually any environmental organization in a major city and the chances are the corridors will be filled with bicycles; if there are any cars parked outside, they will probably belong to visitors, not staff. This has more to do with principle than penury. The car is seen by many environmentalists as a symbol of much that is wrong with our times. In the United States, there are over 155 million on the roads, more than one for every two inhabitants. Add to this the 35 million buses, trucks, and vans that ply 4 million miles of paved road, and the fuel needed to power all these vehicles, and you begin to comprehend the sheer scale of the road transport business. In the United States, there is now more land under roads than under housing, and in the Western world vehicular pollution is the cause of three-quarters of all carbon monoxide emissions, half of all nitrogen oxide emissions, and about one quarter of all carbon dioxide emissions.

All the signs are that the number of cars will continue to rise dramatically, especially in the developing world, where increasing prosperity will inevitably be accompanied by a rise in car ownership.

At present, approximately 65 million new cars are added each year to the stock of around 500 million. Some estimates suggest that the number of cars on the road will double to 1 billion by the year 2030. Whether we like it or not, the car is here to stay. What we need to do—and it does not take an environmentalist to recognize this—is temper its malign influence as far as possible. Currently, many governments seem to be doing precisely the opposite, not least by subsidizing car use and road transportation.

The Appalachians of West Virginia are less than 100 miles as the crow flies from Washington, DC, and their foothills are no more than a two hours' drive away; but they remain a world apart. Midway through the last century, parts of the West were much better known, mapped, and prospected than the Appalachians. Coal and the prospect of wealth and work helped to open up the area, but large tracts remain relatively untrammeled. Drive along the proposed route of Corridor H—or as close as you can to it—from Wardensville in the east to Elkins in the west, and you will traverse a rich variety of landscapes, ranging from gently rolling hills thick with forest, and cleaved by narrow valleys where cattle graze on small farmsteads, to the remote peaks of the Alleghenies. Now and again the two-lane highway dips into a town: Moorefield, surrounded by ugly superstores and pungent-smelling chicken farms; Davis, a popular destination with winter skiers; Parsons, now suffering from the closure of the coal industry. There are few signs of serious wealth but plenty of rural privation. The vernacular architecture is not so much stately white-painted clapboard, although there is that, too, as the ubiquitous trailer. And almost always, parked beside the trailers, are pick-up trucks and the wrecks of old vehicles. Today, the car, not coal, is king in West Virginia.

During the Eisenhower era, a web of inter-states was built across the United States, and the principal ones in West Virginia date from that era. Elkins lies at the heart of a roughly rectangular block of the Appalachians—100 miles across by 200 miles deep—which is bounded by four inter-states: I-79 and I-81 to the west and east, and I-68 and I-64 to the north and south. The four-lane corridors, all of which are identified by letters, were conceived by the Appalachian Regional Commission as a series of parallel roads connecting the inter-states on either side of the mountains.

In one of its earliest incarnations, the 114-mile Corridor H was going to link Elkins with the I-81 in Virginia; during the early 1970s, 6 miles were built to the east of Elkins. Today, this presents a truly strange sight to the visitor on the last leg of the journey from Washington. Heading west from the dramatic bluff of Seneca Rocks, the two-lane highway twists and turns its way across several much-wooded ridges, crosses various tributaries of the Cheat River, then all of a sudden billows out into a four-lane highway with broad verges and a wide median. On the first occasion when I traveled along it—admittedly this was on a Saturday afternoon—only two cars passed me coming the other way, and there were none traveling in the same direction. I was back there on the following Monday, but it was still virtually empty.

The existence of this chunk of Corridor H was the source of some amusement to the anti-Corridor H campaigners. "When I moved here 24 years ago," said Chuck Merritt, who had the looks and demeanor of a prize-fighter, "that stretch of four-lane was under construction. It cost US$75 million in those days, and all it did was go from a lumber mill to a beer joint!" Merritt, whose wife owns the Green House, a handsome building on the outskirts of Elkins that is home to a variety of causes with an environmental slant, sat in on a meeting I had there with Hugh and Ruth Rogers early one Monday.

The Rogers had devoted much of their spare time to campaigning on behalf of Corridor H Alternatives, which opposed Corridor H while pressing for improvements to the existing road system. Hugh Rogers was a lawyer and his wife an artist; both had a quick wit and a nice line in irony.

Hugh Rogers took up Merritt's story in his rich North Carolina accent. "The governor at that time was one of our governors who subsequently went to prison for graft—and he decided that he was impatient with the National Environmental Protection Act (NEPA), and with the federal agencies. He thought, 'We gotta start laying pavement!' So he just rammed it through." Initially the federal agencies remained mute; but when the construction work cut off the water supply to a federally funded fish hatchery, they swiftly intervened and construction was halted.

By the early 1980s, the Corridor H project appeared to be dead. There were no federal funds available. However, by the end of the decade it was resurrected, not least because the political portents were in its favor. Robert C. Byrd, one of the state's senators and a fervent champion of development, had become chairman of the Appropriations Committee in the Senate. This is the closest any politician gets to being handed the keys to the Treasury safe. The West Virginia Division of Highways, always a keen supporter of Corridor H, commissioned a new environmental impact study for a variety of routes through the Appalachians. Following public hearings in 1993, the Division of Highways eventually opted for a scheme that would create a four-lane highway through the mountains some way to the north of the original route, passing beside or through Kerens, Parsons, Thomas, Davis, Moorefield, and Wardensville.

During the early years, the sole justification for building Corridor H was economic. A new highway, it was argued, would open up the region to new business opportunities, and this argument was still forcibly advanced by representatives of the Corridor H Action Committee, who requested that I meet them in the none too salubrious surroundings of the Elkins branch of McDonald's. I arrived a few minutes late and found Jim Kingsbury, founder and chairman of what he called "our working-class orientated group", and Charlie Phillips sitting in a corner of the restaurant, the table in front of them submerged beneath a great weight of paperwork. Kingsbury wore glasses and a pinstripe suit. He had a plump face, a quiet voice, and a very still manner. He gave the impression of a snake waiting to pounce on its unfortunate prey. Phillips was a good deal older, probably in his sixties, and his craggy features and worn hands suggested he was no stranger to manual labor, although I notice that in the organization's literature he describes himself as a "political and social commentator". As we shook hands, Kingsbury asked if I had done as he had suggested in an earlier phone conversation. Yes, I replied, I had spent Sunday morning driving along the proposed route of Corridor H, from Elkins up to Parsons and Thomas in Tucker County.

"And how was it?" he asked. I told him that the highway had been carpeted by a soft blanket of fresh snow, and at one point I even thought of turning back for fear of getting stuck. "You can see why we need the new road then," said Kingsbury emphatically. As it happened, Route 219 from Elkins to Thomas had struck me as being a perfectly good road, at least by European standards, though I might have thought differently had I been stuck behind a lumber truck on the steep hills, a common enough complaint around these parts.

"In Tucker County," said Kingsbury, "you have a very depressed economy. It was coal-mining based and it still should be, but the mines have gone down mainly because of environmental regulations. What Tucker County needs is economic development. It needs more jobs and Corridor H is the one hope that they have in that area." During recent months, explained Kingsbury, the authorities had been searching for a site in order to build a new regional jail. Parsons had hoped it would be chosen because the jail would eventually provide 150 new jobs, in addition to immediate construction work. In the end, it was struck off the list of candidates. "Had Corridor H been completed, they might have had a better chance," surmised Kingsbury.

While Kingsbury was describing the gloomy work prospects for young people in Parsons, Eddie Canterbury arrived. A small dark-haired man with bony features and a didactic, argumentative manner, Canterbury had spent over 30 years working for the US Forest Service before being elected as a county commissioner. "It's an established economic fact," he said, "that economic development doesn't occur on two-lane highways. It will only occur on four-lane highways." As far as he was concerned, Corridor H should have been built 20 years ago, and had it been it would have cost 40 percent less than it will now. The environmentalists, he added, should take the blame for this fiscal impropriety.

I noticed that as time passed, Kingsbury became increasingly restless, and he kept looking over his shoulder. Was he, I wondered, expecting someone? Yes, he was. He had telephoned a television station called Channel 6 and he was expecting them to come along to film "our little meeting". I decided that it was time to take my leave, but before I could do so a woman of Chinese origin who had been sitting at a table nearby, and seemed to have something to do with the television station, came outside at Kingsbury's request and took a photograph of us all. Strange.

The views from Beander's bar echoed those of the Corridor H Action Committee. Without the road, the region would remain economically backward; build Corridor H, and jobs and prosperity would surely follow. "My family owns a saw mill in the hills," explained Beander. "One year the weather was so bad that the trucks refused to collect the lumber." Had there been a good road, he said, they could have reached his mill.

Propping up the bar was a pleasant young man from Pittsburgh with a fashionable layered haircut and some serious tattoos. He had recently been posted to Elkins as manager of a new fast-food restaurant. He thought the proposed highway was a necessity, too. Why? "Before we opened we had 650 applicants for 84 jobs," he replied. "We are not talking about good wages for skilled jobs. We are talking US$5.15 an hour minimum wage. That just shows you how bad people want work round here. The highway'd be bound to help."

Beander's mother, Eunice, a trim, hawk-eyed woman with gray hair and fancy spectacles, said that all the businesses in Elkins—the people who provided jobs—were in favor of the road. "You should've been here when they had the rally," she added, "141 rigs

went down this street, blowing their horns. It was mighty impressive!" Another rally was planned for Moorefield in two weeks' time.
Its sponsor was the Committee for Corridor H, which represented
business interests in the area. At least 200 trucks were expected to
parade past the governor, Cecil Underwood, two state senators, and
various county officials. Among the bands scheduled to perform
at the rally was the aptly named Lonesome Highway.

However, not everyone was impressed, either by the rallies or by
the economic arguments being put forward by the pro-road lobby.
"I think we surprised them by taking on the economic argument,"
said Hugh Rogers. "For them it was an article of faith: roads bring
economic development." He and his colleagues at Corridor H
Alternatives in Elkins teamed up with campaigners at the other
end of the proposed road in Wardensville in 1993, and they
immediately began to challenge the economic arguments. Evidence
from earlier corridors suggested that in sparsely populated rural
areas jobs were more likely to be lost than gained when new four-
lane highways were built. They found an important ally in Terance
Rephann, who worked on road research in the West Virginia
Regional Research Institute. "He wanted the truth," recalled Rogers,
"and it turned out to be a very hot thing." Rephann's studies, funded
in part by the corridor promoters, the Appalachian Regional
Commission (ARC), and by the federal government, challenged
the saloon bar wisdom. Rephann found little evidence that new
highways stimulated economic development in rural areas. In fact,
the opposite held true. "New highways usually result in either no
net change or accelerated decline of rural communities," he wrote.
"Regions which are already relatively urbanized, and situated in
close proximity to other major urban areas, experience more growth
than isolated rural regions."

Rogers had been thorough in his research, and among those he
had talked to was the first executive director of the ARC, who
believed that on economic grounds Corridor H was the least
justifiable of all of the corridors. He told Rogers: "It was a stupid

idea. I didn't want to build it." This was a view shared by the US Environmental Protection Agency (EPA), which in 1996 declared that Corridor H should not be built for two reasons. Firstly, it received the lowest environmental rating a highway could receive; in other words, it was about as damaging as it could possibly be. If built, Corridor H would do significant damage to two national forests, 41 streams, considerable expanses of farmland, and many celebrated historic sites, including two civil war battlefields. Secondly, the EPA found that there was no evidence that Corridor H would yield any significant economic benefits for the region. Carol Browner, the administrator of the EPA, was called in front of a congressional committee and berated by West Virginia's two senators, Robert Byrd and Jay Rockefeller. Suitably chastened, Browner then ordered Peter Kostmayer, regional director of the EPA, to change his report. He refused and was subsequently fired from his job.

Instead of engaging the campaigners at Corridor H Alternatives in rational debate on the economic issues, the pro-road lobby had done its best to denigrate them. The road lobby's arguments against Corridor H Alternatives were always, as Rogers put it, *ad hominem*: they attacked the individuals, not the ideas. Charlie Phillips was certainly robust in his condemnation, and when I told him that I had seen Rogers and Merritt he dismissed them as "troublemakers off the street." Rogers actually looked more like a Wall Street banker than some crack-crazed anti-capitalism protester, but this didn't impress Phillips. He had been equally forthright in an article that he had written for a local newspaper. "Who caused the closure of the coal mines, steel mills, power plants, timber industries, mom-and-pop businesses, and so forth?" he demanded. "Those who caused this are none other than the pantheistic radical environmentalists, the same people who made environmentalism their 1990s substitute for the 1960s protests against the Vietnam War. I say this to the environmentalists: get over it! Get a life and move on."

Well, I did move on, heading to Wardensville to meet the main representative of Corridor H Alternatives at what will be Corridor

H's easterly starting point if it is ever built. Bonni McKeown looked just like the sort of person who would have opposed the Vietnam War and shouted subversive slogans on the street. She had a beaten-up Subaru covered with stickers promoting causes such as organic farming and kindness to animals, and in the back was a rucksack, an unfurled sleeping bag, and the detritus of campaigning: leaflets, papers, maps, and the like. She wore baggy pants, steel-rimmed spectacles, and she laughed a lot as she took me on a brief tour of her home town. There was a small one-room jail, dating from before the Civil War, and a street of handsome wood-framed houses dating from the 19th century. A herd of cattle grazed in a meadow on the edge of the main street and flower-speckled fields undulated up to a tree-clad hillside. If a rural idyll existed in West Virginia, this was it. And the idyll certainly won't survive if Corridor H is built.

McKeown's opposition to Corridor H had come at a considerable personal cost—she had lost her job and been vilified in the press—and this went some way towards explaining why she declined to enter a restaurant with a "Corridor H Finish It!" sticker on its window. Instead, we went to a cavernous bar where the only other customers were five men in baseball caps who were hunched morosely over their beers like bilious vultures digesting a poor meal.

In McKeown's view, the economic arguments for Corridor H simply didn't stack up, and the Division of Highways' own traffic forecasts conclusively proved that a 100-mile four-lane highway was not needed. Projected traffic on existing east–west roads was likely to average less than 5000 vehicles a day for the next 20 years, or around half of what engineers generally consider to be the traffic volume that justifies a four-lane highway. Now that the environmentalists had highlighted the flaws in the road-equals-jobs-and-prosperity line of thinking, the road-building lobby sought other arguments to justify the spending of around US$1.3–US$1.5 billion of taxpayers' money. Road safety, scarcely mentioned 30 years ago, had become a key issue, and Jim Kingsbury and his colleagues went so far as to suggest that the environmentalists, by opposing Corridor

H, were responsible for the deaths caused by the present poor road conditions.

McKeown conceded that in some places the road across the Appalachians was dangerous, and Corridor H Alternatives had devised a scheme that would iron out dangerous corners, add passing lanes in places and make other improvements. They agreed that in a few areas where there was heavier traffic, three- or four-lane sections might be justified. "Of course, we are not engineers," admitted McKeown, "and we've had to rely on the Division of Highways' figures for traffic projections." It was a pity, she suggested, that the highway agencies had spent millions of dollars on fancy reports describing their plans for Corridor H and detailing their environmental impact, but had done nothing to evaluate the sort of improvement schemes she and her colleagues were promoting. The Division of Highways, she added, was also spending—illegally— tens of thousands of taxpayers' dollars setting up a website devoted solely to promoting Corridor H.

But how much would Corridor H Alternatives' improvement scheme cost? "Well," said McKeown after some consideration, "it costs around US$2 million to correct a dangerous curve. And you should compare that to the US$15 million it costs to build each mile of new four-lane highway." Chuck Merritt was also forced to speculate. "If I had to pull a figure out of my behind," he said, "it would be about the amount of money they have now. They could do a bang-up job for that." Aware of the fact that their opponents could point to their lack of expertise, Corridor H Alternatives was in the process of hiring an engineer to cost out their plans for road improvements.

The 1998 Federal Transportation Funding Bill—TEA-21—and the appropriations bill that followed it authorized around US$2.55 billion for the entire Appalachian Corridor system in 13 states. West Virginia's share of this was around US$345 million. Even if this were devoted solely to Corridor H, it would not be enough to build a quarter of the highway.

The building of the latest version of Corridor H should have begun by 1995; but Corridor H Alternatives sued the West Virginia Division of Highways and the Federal Highway Administration for violating the National Environmental Protection Act by refusing to consider the sort of alternatives that they had been promoting, and failing to determine the precise impact of construction on historic sites. The highway agencies won, but the environmentalists appealed, and the legal tussle had kept the road at bay. During the time of my visit, the Division of Highways was evaluating in greater detail, at the US Court of Appeal's direction, the environmental impact of Corridor H on historic sites. In 1998 work had begun on a 3.5-mile by-pass around Elkins—the environmentalists had no objection to this—but in early 1999 the heavy equipment brought in to start construction of the corridor east of Elkins was lying idle in a field beside Route 219.

"You know something?" asked Beander. "It's costing the taxpayer US$46,000 a day in rent with all that equipment lying idle." And that, he said, was the fault of the environmentalists. When I repeated Beander's assertion to Hugh Rogers, he was dismissive. The Division of Highways, he said, had been well aware that there was the possibility of an injunction, yet they still put the work out to tender and behaved as though it was definitely going to proceed.

*   *   *

Some subsidies are relatively easy to identify and measure, and their impacts are quantifiable, too. Take, for example, the Alaska logging subsidies. We know almost to the dollar how much taxpayer money has been channeled into the state, who the beneficiaries have been, and the impact the subsidies have had upon the forests. It is somewhat harder to assess subsidies to the livestock industry. However, it is still possible to come up with a reasonably accurate figure for some subsidies—for example, for predator control in the United States—and it is possible to identify who the main beneficiaries are, and what effect these subsidies are having on the environment and on trade. By comparison, transport subsidies are ferociously complex, and while it may be relatively easy to come up with a critique of a specific road scheme—especially one so expensive and damaging as West Virginia's Corridor H—putting figures to the overall subsidy, in the United States or elsewhere, is far harder.

Public expenditure on road transport takes a variety of forms, from the building of roads and related infrastructure to their maintenance and repair, and the provision of traffic control and emergency services. But governments also impose charges upon road users, most obviously through vehicle taxes, license fees, tolls, and road pricing. The subsidy to the automobile is therefore the difference between public expenditure on road transport and receipts from users. The precise quantity of subsidy obviously depends upon what you choose to include. If you factor in such things as tax breaks for business cars, the failure to charge a market rate for city parking, and the tax loopholes that mean that sports utility vehicles are not subject to the fuel-efficiency taxes that are imposed on other cars, then the subsidy will obviously be higher.

In *Addicted to Subsidies*, Cees van Beers and André de Moor estimate that governments around the world spend US$225 billion a year on automobile subsidies. Over half of this—some US$125 billion—is the annual subsidy in the United States, which is equivalent to around US$660 per vehicle. This compares with US$567 per vehicle in Japan and US$355 in the UK. In France, on the other hand, fuel and other taxes are so high that road users actually pay the government an implicit tax of US$172 per vehicle, or around one fifth more than the share they would pay if they financed the direct costs of road use.

Two studies carried out in the United States—one by the World Resources Institute (WRI), the other commissioned by the Organi-sation for Economic Co-operation and Development (OECD)—came up with annual subsidy figures for US road transportation of US$174 billion and US$55 billion respectively. While the former factored free employee parking into its calculation, the latter did not, and this helps to account for the large gap between the two figures. However, whichever figure you care to take, American motorists still only pay a fraction of the actual costs of their travel: 20 percent if you take the higher figure, 50 percent if you take the lower figure.

If we decide that resource users—in this case, the owners of automobiles, vans, trucks, and so on—should pay the full costs of road transportation, should we include, in addition to the direct costs (the building of roads, traffic control, highway policing), the external costs of road use that drivers impose upon society and the environment? The most obvious of these is pollution, which affects air, water, crop productivity, wildlife, climate, and human health. Ideally, we should, but how does one measure the costs? When low-lying islands in the Pacific disappear—and global warming may lead to a sufficient rise in sea level to ensure that some do—how will we calculate the economic costs of such a loss. How would we allocate a share to car drivers, and which car drivers in which countries would be culpable? And here is another conundrum. The British eco-pundit, Norman Myers, suggests that road users in the United States should pay some of the military costs of protecting oil shipping routes in the troubled Persian Gulf. Over half of the oil imported to the United States from the gulf is used for road transport; therefore, motorists, says Myers, should pay half of the US$50 billion military expenditure bill.

However, just because we are unable to attach precise figures to the environmental externalities does not mean that we should ignore them. On the contrary, governments, and for that matter communities and individuals, should do all they can to reduce the negative impact of car use. Even those who are obsessed with cars cannot deny that many cities and towns, and huge expanses of countryside, have been ruined by roads and traffic; that vehicular pollution is a serious problem, both locally in some places, and globally, especially in terms of its contribution to global warming; and that road transport relies on the use of non-renewable resources, such as oil. It is in all our interests to use less, not more. The very least we should do, therefore, is insist that drivers pay the full, direct costs of driving. If they did so, then they would, for the most part, use their vehicles more sparingly. In any case, why should those who do not use cars subsidize those who do? This is an equity issue, too.

In the United States, successive governments have deliberately sought to promote private transport and road-building at the expense of public transport and the use of rail and street cars. Needless to say, the automobile and oil corporations have connived in this battle to make the car supreme. The road-building and automobile lobbies are immensely powerful, and through a combination of persuasion and carefully targeted political campaign contributions they have been able to ensure that government policies work, by and large, to their advantage. It seems that no mainstream politician in the United States would dare to campaign on an anti-car platform, or even to suggest that it would make better economic and environmental sense if vehicle users paid the full cost of vehicle use. It is depressing to report that the federal government has recently increased spending on transport, instead of reducing it. Capitol Hill appears to be oblivious to the problems caused by cars and other forms of road transport. The 1998 Transport Equity Act for the 21st century—TEA-21—guaranteed a 47 percent increase in highway funding over the following five years, and a good chunk of this is "pure pork". It will specifically be spent on projects that have been requested by individual members of Congress for their own regions. These schemes are often known as demonstration projects. In 1982, there were ten demonstration projects costing US$362 million; the 1998 TEA-21 authorized over 1850 demonstration projects at a cost of US$9 billion.

Almost double this sum will be spent if the 50 roads highlighted by *Road to Ruin* go ahead as planned. Some are relatively modest, but others, such as Corridor H, will cost the taxpayer a fortune and provide few of the benefits that their proponents claim for them.

Many environmentalists believe the laws governing road-building are far too lax; if they were subject to proper cost–benefit analyses many of these schemes would never even be considered. "The real problem," suggested McKeown, "is that the law is so weak. There are no real laws preventing bad projects of this sort. At least with things like dam-building, the Army Corps of Engineers has to do a cost–benefit analysis. With roads, no cost–benefit analysis is required, and we are forced into environmental and historical nit-picking."

Let us assume, for the sake of argument, that Corridor H and the 49 other road projects identified by *Road to Ruin* as fiscally imprudent and environmentally harmful are shelved. That would save the US taxpayer US$17 billion, prevent the destruction of many fine landscapes and much wildlife, and protect scores of small towns and communities from the blight of traffic. Let us hope—some hope! —that this will happen; but even if it does, it will do little to reduce the miles driven each year (2 trillion miles in the United States) and the amount of car-related pollution that is pumped into the atmosphere. We need to do far more than cancel road schemes if we are to prevent the car from further despoiling the planet.

There is some evidence to suggest that measures such as road pricing and increasing gasoline taxes can have an effect. In Singapore, the introduction of user fees for cars entering the city center helped to reduce peak-hour traffic by 75 percent. This, in turn, helped to reduce the amount of vehicle pollution. The pro-car lobby invariably claims that road tolls have an adverse effect on the economy. However, this ignores the obvious fact that while a US$5 toll will deter private car owners from making a short journey to buy a 60-cent loaf of bread, it is an insignificant charge for a truck delivering US$20,000-worth of goods. However, introducing what many see as punitive measures in an authoritarian state such as Singapore is one thing; doing the same in a liberal democracy is quite another.

When the oil crisis bit during the 1970s, and gas prices soared, many Americans sold their gas-guzzlers and bought smaller, more fuel-efficient cars. Unfortunately, when prices dipped in the 1980s, they went back to the gas-guzzling, sports-utility type vehicles, and while gas prices remained low there was no financial incentive for Americans to opt for smaller, less polluting, more fuel-efficient cars. When gas prices shot up in late 1999 and the early months of 2000, there was much squealing and grumbling from road users, but no appreciable change in purchasing behavior. A Sierra Club energy expert told me that he had expected new buyers to opt for gas-efficient small cars, rather than gas-guzzling, sports-utility vehicles, but the latter were still selling in vast numbers. This suggests that if governments are to encourage drivers to use more fuel-efficient cars, gas taxes either have to rise by a considerable amount—far more than most people will accept—or other measures, such as differential taxation to encourage fuel efficiency, will be required.

Of course, this is an exceptionally complex issue and transport policy cannot be conceived without reference to a whole range of other factors, such as housing, industry, and demographic trends. What is good for one region may not be good for another. Nevertheless, those experts who have worked on this vexatious issue are agreed that certain reforms are essential. Most significantly, road users should pay the full costs of road use, and, ideally, nations should act in concert.

*   *   *

Let us get back to the issue of democracy, so eagerly and, some would say, eccentrically championed by Jim Kingsbury and his colleagues. Those who are in favor of Corridor H make much of the fact that they are a majority, and that in a democracy their views should hold sway. The Committee for Corridor H cites a poll that found 77 percent of residents in favor of the road as part justification for its construction. The West Virginia Division of Highways' web site devotes considerable space to analyzing public opinion, as though this in itself added weight to its case. And Jim Kingsbury and the Corridor H Action Committee made much of the fact that the majority of people who were living near the proposed road were in favor of it. The great pile of paperwork that Kingsbury brought with him to McDonald's was a petition calling for Corridor H to be built. Organized by his committee, it had been signed by 10,500 people: "Mainly working-class people, plus some business people, elected officials, and the governor, Cecil Underwood," explained Kingsbury.

This failed to impress Bonni McKeown, who claimed that many of those who signed were from places that were far from the route of Corridor H. She also thought that many had been inveigled into signing by local politicians and activists. In any case, when there was a public consultation process in 1995, almost 90 percent out of the 4000 comments received by the state were from people who opposed Corridor H. This, in Hugh Rogers's view, was "a poll of people paying attention."

Kingsbury saw it differently: "What you have there is a galvanized, energized minority that does express their views," he said. "The working class by and large don't become that activated. That's why we did the petition." Besides, said Kingsbury, one newspaper poll found 88 percent of local people in favor of Corridor H. If democracy was working properly, then it would be built. However, this ignores the fact that Corridor H is not a local issue. It is a national issue, and if it were built it would be largely funded by US taxpayers, the vast majority of whom would oppose the scheme if they were in possession of the facts.

When I asked Eddie Canterbury whether he ever worried about the high costs of constructing Corridor H, he replied: "No, sir. I really don't. I've been working on this road for 20 years. I worked on teams that studied routes through the National Forest. For those 20 years this part of West Virginia did not grow or prosper one iota because there's no highways, no access." According to Jim Kingsbury, Corridor H was no more than the state deserved. Opponents of the road, he suggested, should remember this: "The state has contributed to the nation as a whole through coal mining. It's not as though we are asking for something we haven't contributed many times over."

This theme—that a state was somehow owed something for the goods and services it had provided in the past—was one I heard frequently on my travels. And, of course, it is a game that every state can play. California might cite gold; Florida, sugar; Texas, oil; Oregon, timber; Kentucky, tobacco; Colorado, water; and so on. As a means of determining how financial resources should be allocated, it simply doesn't bear examination. It may be possible to argue that the extraction of coal in West Virginia did the state more harm than good. In environmental terms, that is undoubtedly the case, and coal made vast fortunes for a few, while providing tens of thousands with little more than a meager living and lung disease. However, the idea that taxpayers from California or Colorado should, in some way, compensate the people of West Virginia for this is plainly ludicrous.

It seems that the engineers of West Virginia Division of Highways were just as cavalier about spending taxpayers' dollars as Eddie Canterbury, and Ruth Rogers recalled a bizarre meeting in the state capital, Charleston. Among those present were the director of the Division of Highways, officials from the Federal Highways Administration, and a woman from the Advisory Council on Historic Preservation. At one point during the meeting, the highway engineers said that they had come up with a plan to avoid damaging a famous civil war battlefield. They would build "on structure" and construct 6 miles of Corridor H on stilts, skirting around the battlefield. Ruth Rogers expressed surprise. "Oh, it's really easy," they told her. "It'll be just like the Linn Cove Viaduct on the Blue Ridge Parkway in North Carolina." Ruth pointed out that the Linn Cove Viaduct was a two-lane, with a speed limit of 45 miles per hour. They said: "No, it's not, it's a four-lane." Ruth explained that she came from North Carolina and she *knew* that it was two-lane. Eventually they accepted what she said. Still, it wouldn't be a problem, they insisted. They would build a four-lane version.

When Ruth got home she told her husband about the meeting and he called up the Florida engineers who had built the Linn Cove Viaduct to find out how much it had cost. "They rang me back after a while," recalled Hugh Rogers, "and the first thing they said was: 'You've got to remember the costs are from 16 years ago, so you'll need to factor in inflation.' And then they gave me a quote—not per mile, but per square foot! It was that expensive!" At 1980 prices, a two-lane viaduct cost around US$33 million a mile. Let us say that you add another 50 percent for two more lanes. You will come up with a cost of around US$50 million per mile. At that price, 6 miles of viaduct would work out at around US$300 million, and then you need to factor in inflation. Even without taking inflation into account, this figure is close to the entire amount of federal money allocated for West Virginia's entire corridor system over five years. Clearly, the engineers promoting the scheme had never given a moment's thought to the cost. And let us remember, most of this money is the US taxpayers', not West Virginia's, although the state would contribute one fifth of the costs.

Some time after I left West Virginia I found myself in Michigan, chatting to a transport campaigner who worked for the Michigan Land Use Institute. I mentioned Corridor H, and he said: "That's strange." He opened up his e-mails and printed one that had arrived that morning. It was a press release from Corridor H Alternatives announcing that the lawsuit had been settled. "As a single project, Corridor H is dead," was its heading.

That night I rang Hugh Rogers and he told me what had happened. Corridor H Alternatives and its 14 coplaintiffs had reached a settlement in court with the relevant state and federal agencies. The proposed highway had been broken into nine sections. Some would go ahead soon; others would not be built for many years, and possibly not at all if anti-road resistance is successful. In some areas, the highway department had agreed to draw up entirely new plans in order to safeguard important historic sites, and east of

McKeown's home town of Wardensville, the highway department had agreed to a 20-year delay.

It does not require much courage to work for an environmental organization in Washington, DC, or in any other city. In fact, I often felt that many of the metropolitan greens I met led very comfortable lives. They move in a society broadly sympathetic to their beliefs, campaigning against activities and businesses from which they are often far removed. But it is a very different story in places such as rural West Virginia. Here, McKeown and her colleagues walk the same streets as their opponents, and frequently find themselves being openly vilified and harassed. Their courage and good humor are admirable. The settlement was by no means an unmitigated triumph for the anti-road lobby; but considering the might of their opponents, they had done exceptionally well to get the concessions they did.

If Corridor H or substantial parts of it are eventually built, some sections will probably be named after Senator Robert C. Byrd, one of the scheme's most fervent advocates. "There's a joke round here," said McKeown, "that there are now so many highways with his name that if you give someone directions to the Robert C. Byrd highway, they'll be driving all over the state." A few years ago, McKeown and Hugh Rogers went to Washington, DC, to discuss Corridor H with the senator. "He's in his eighties now," recalled Rogers, "and he kept nodding off. But one thing got his attention. I said: 'You're an historian. You know Lincoln's second inaugural address—as our times are different, so must our policies be different.' Well, he woke up, slammed the table with his fist and shouted, 'Our times are not different!'"

Rogers had a sneaking admiration for the senator, and he stressed that while the state had a long and dishonorable tradition of electing governors who eventually ended up in prison for corruption, there was never any suggestion that Byrd was anything other than honest. He simply saw West Virginia as his fiefdom, and in the eyes of many West Virginians his success as a politician, as the democratically elected overlord, could be gauged by the amount of federal dollars he squeezed out of Washington. Over the years, he had proved himself a winner. For him, times had not changed. The wilderness needed to be tamed when he first entered the House of Representatives in 1953. It still did, and he frequently quoted the verse from Isaiah that I mentioned in Chapter 4: the one about making the rough places plain, straightening the crooked and making straight a highway for our God. This verse provided a biblical legitimacy for the building of Corridor H, and its opponents could go to hell.

CHAPTER 8

MINING THE TREASURY

I had seen pictures of the abandoned Summitville gold mine. I had read endless facts and figures about the damage that had been done to this remote region in southern Colorado, and about how a mining company had carved almost 1000 acres of soil and rock out of a craggy mountainside in its quest to extract gold. But nothing had prepared me for the sheer scale of the mine workings. Driving up to the mine site, after a long uphill haul through dense pine forests and alpine meadows, we must have looked like Lilliputians entering a giant's world.

Ignacio Rodriguez and I parked beside a mud track and made our way through fresh puddles of water to a small hut that was used as an office. We signed ourselves in, donned hard hats, and left for a tour of the site with Camille Price, an employee of the state of Colorado, in her mud-spattered four-by-four vehicle. Price was friendly enough, but she was clearly eager to hurry us round the site. Providing a guided tour was not a priority. We drove up one side of a great hillside of spoil, past dirt-moving trucks the size of two-story buildings, then skirted around a lagoon of greeny-blue water. All the while Price ran us through the history of Summitville.

During the 19th century, she explained, thousands of miners headed through Colorado on their way to join the San Francisco gold rush. Some made a mental note of the fact that there was gold here, and they returned during the 1870s. Working either alone or in small groups, they followed the underground veins; over the next century numerous shafts were dug. The legacy of these tough and no doubt rowdy times could still be seen in the derelict wooden shacks where the miners lived, and which clung to the hillside above the lagoon.

In 1986, the whole way of doing business changed with the arrival of the Summitville Consolidated Mining Company, which was owned by Gallactic Resources. Instead of digging for gold in the old manner—a process that was conducted underground and left little on the surface other than mine tailings—the company set about extracting gold using a technique known as cyanide heap leach mining. This is chemistry practiced on a grand scale. Crushed ore is dumped on a supposedly impenetrable plastic pad and cyanide is sprinkled over the ore. The cyanide extracts the gold and other precious metals.

"But things didn't go as planned," said Price with admirable understatement. Cyanide leaked through the plastic pad, which had been ineptly installed during the winter months, and the sheer scale of the operation meant that the natural process of acid rock leaching was greatly hastened. A combination of cyanide and sulfuric acid killed all life in the Alamosa River, which to this day has a pH not much higher than vinegar. Attempts by the mining company to sort out the pollution failed. Eventually, Gallactic Resources declared bankruptcy and ran. Its Canadian owner, Robert Friedland—or "Toxic Bob", as he has been dubbed by environmentalists—continues to operate in many other parts of the world, while the US authorities do their best to clean up the mess he left behind.

In December 1992, Summitville was declared a Superfund site and the US Environmental Protection Agency (EPA) began the task of stemming the pollution flowing from the mine. Since then the clean-up has cost around US$40,000 a day, and will continue to do so until the Alamosa is returned to its original state. A hefty chunk of Superfund money comes from taxes paid by oil, mining, chemical, and other companies that produce, mine or refine potential pollutants. However, taxpayers get stung too. For the financial year 2000, almost half of the US$1.5 billion allocation for Superfund clean-up operations came from general taxpayer funds.

When I first spoke to Rodriguez on the phone, I understood him to be a rancher. In a small way, he had been, but only after he retired. A native of Texas, he had worked for many years in mental health administration, and during his vacations he and his friends often came to these mountains to hunt and fish. It was these experiences that encouraged him to retire to Colorado. Tall, sinewy, and elegantly dressed in a white straw hat and pressed denim shirt, he spoke impeccable English with a Hispanic lilt, and his actions and thoughts were imbued with great dignity. He had hoped, on retirement, to spend his time tending the few cattle he had, fishing the Alamosa, which runs through his land, and busying himself around his small ranch. Summitville changed all that, and he had devoted much of his retirement to voluntarily working with the Technical Assistance Grant (TAG) group that had been set up to provide a forum for citizens concerned about the Summitville clean-up. The TAG group was trying to ensure that the authorities made a good job of it. He was, by and large, unimpressed.

"The EPA did well up to a point," he explained as we headed back down the mountain to meet some fellow members of the TAG group. "Initially, they were very aggressive in setting up a plan to clean up the site." However, until very recently they had ignored the advice of locals such as him, and of the expert witnesses hired by the TAG group. "They've spent something like US$150 million," he said, "and we reckon they could have achieved the same for US$90 million." One of the EPA's most notable blunders was to block an adit, against the advice of the TAG group and their hired experts. "All they did by doing that," reflected Rodriguez, "was turn source pollution into non-source pollution." Instead of polluted water being channeled into one place, it was being diffused around it. "They just wouldn't listen to us," he recalled. However, over the past year relations between the local citizens and the EPA had improved; but Rodriguez was still scathing about the role that the state had played in all this, as was Wendy Mallot, whom he described, as we pulled into the yard outside her home, as one of the spark plugs of the group.

Wendy Mallot's house was situated a few yards from a rail track in a lush valley hemmed in by dense woodland. There were hunting dogs penned up outside, and in the living room the skin of a mountain lion was tacked onto the wall. It was clear that Mallot knew the mining business backwards. Her father had been the foreman of an underground gold mine at Summitville until 1969, and she had worked there as office manager from 1987 to 1990. Now, she was secretary of the TAG group. "It's been frustrating," she suggested at one point, "but the group has made a great deal of difference. Whether they like it or not, we're going to be heard. And the Alamosa is not going to be clean until the citizens say it's clean."

Mallot felt that the local communities had been betrayed by the state. Before they became involved in the Summitville clean-up, they were under the impression that the Division of Minerals and Geology—the department that grants permission to mining companies—acted as a watchdog. Instead, they discovered that it was actively promoting mineral development. Right from the start, Gallactic Resources was allowed to do things that the state should have prevented. Most obviously, it failed to lay down the plastic pad in a way that would prevent leakage of cyanide.

Throughout my day with Rodriguez, he kept referring to the 1872 Mining Law. It was his belief that the law had enabled companies such as Gallactic Resources to plunder the planet, make serious profits, and pay nothing for resources that belonged to the American public. For the past 130 years, he explained, companies have been able to extract gold, silver, copper, and other hard-rock minerals, from public lands without paying either a fair price for the land, or any royalty on the minerals extracted. According to the Mineral Policy Center, royalty-free mining on public land had handed the mining industry a subsidy worth some US$230 billion.

*   *   *

I had been pointed in the direction of Ignacio Rodriguez and Summitville by Aimee Boulanger, whom I had met some weeks earlier in Montana. From a modest office in Bozeman, she acted as a quaintly termed circuit rider for the Mineral Policy Center, an organization which has been campaigning for reforms of the 1872 Mining Law since its inception in the early 1990s. Although circuit riders—there was another circuit rider for the southern Rockies whom I was to meet shortly—spend much of their time working with people such as Ignacio Rodriguez, they must also face up to the mining companies, to their employees, and to organizations such as People for the USA. This particular organization frequently takes great exception to those who challenge the right of mining companies to dig for gold and other minerals. When I dropped by the Bozeman office, I expected to find a solid-looking woman in jeans, boots, and shapeless sweater, with the sort of demeanor you might expect of a woman who spends her time in a man's world and doesn't want her femininity to get in the way. Instead, I was confronted by a young woman whose elegance was matched by her wit and intelligence.

"We don't object to hard-rock mining," she said as soon as I sat down, "providing it can be done decently." There was one caveat to this. The Mineral Policy Center believes that there are some places—pristine wilderness and natural landscapes—that should not be subject to mining activity under any circumstances.

Much of Boulanger's time is spent traveling around the West, visiting communities who are concerned about specific mining activities and who have requested her help. Many have problems that relate to existing or abandoned mines, while some are concerned about the impacts of planned mines. In Boulanger's view, many of the problems that communities face stem from the weak legislation governing hard-rock mining in the United States. In 1872, President Ulysses Grant signed the General Mining Law, one purpose of which was to encourage the settlement of the West. It was, said Boulanger, a law whose time had come and gone, and it was urgently in need of reform.

Under the law, hard-rock mining is deemed "the highest and best use of public lands." In practice, this means that public land managers such as the Bureau of Land Management and the Forest Service are obliged to give mining claims priority over all other uses, whether for recreation, timber extraction, grazing, or wildlife management. At present, anyone who discovers a valuable mineral deposit on public lands has a right to mine it. "What we're calling for is discretion," explained Boulanger. "At the moment, public land managers cannot say: 'No, we think this—recreation, conservation, whatever—is a better use of the land than mining.'"

This lack of discretion had recently led, ludicrously, to President Clinton spending US$65 million of public money to buy out the rights of a company to mine on public land just outside of Yellowstone National Park. Had the public land managers weighed up mining against other interests, they would undoubtedly have rejected the company's bid to mine this environmentally and aesthetically important area.

The Mineral Policy Center's opposition to the Mining Law was shared by no less a figure than Mike Dombeck, then chief of the Forest Service. When Dombeck resigned in March 2001—prompted, many believe, by unhappiness with the Bush administration's plans for his department—he wrote in his letter of resignation: "Every single use of the National Forest System—recreation, timber harvest, oil and gas development, for example—is subject to the approval or rejection of a field official for environmental or safety reasons. All but one—that is, hard-rock mining." He then called on Congress to reform this "anachronistic law."

The 1872 Mining Law introduced the concept of patenting on public land. Until recently, anyone who could prove the existence of valuable mineral deposits, and show that they had the means and wherewithal to exploit them, could patent, or purchase, the land and minerals for US$5 or less an acre. In 1994, Congress passed a law that introduced a one-year moratorium on patenting; every year since then the moratorium has been renewed for a further year. None of this has made any odds. Mining corporations are still extracting minerals on public land and paying nothing for them. "Instead of having annual moratoriums," suggested Boulanger, "we need to do away with the whole concept of patenting. No other industry can privatize public lands as the mining industry can. It's simply a wholesale give-away."

And the same could be said when it comes to royalties, or the lack of them. Oil and gas companies operating on public land in the United States pay a 12.5 percent royalty; coal companies pay 8 percent; hard-rock mining companies pay nothing at all. "It's ridiculous," said Boulanger. "Do we think that strategically gold is more important than oil or gas or coal, and that we should be subsidizing companies to get the stuff out of the ground?"

Mark Twain once described a mine as a hole in the ground with a swindler on one side and a sucker on the other. If this chapter has more to say about the suckers, the people who suffer from mining, than their adversaries, the mining companies, then it is largely because the latter proved hard to contact. Gallactic Resources, who presided over the Summitville disaster, had long since ceased trading in the US, as had its erstwhile owner, Robert Friedland. Battle Mountain Gold, whose activities I was to inspect after I left Summitville, had a few laborers on site still at its mine near San Luis, but no managerial staff. I fared no better when I headed down to Silver City, New Mexico, a town whose wealth has much to do with the vast copper operations run by Phelps Dodge. I failed, despite plenty of attempts, to get an interview with the managers of the mines.

However, when I was in Washington, DC, I did get to visit Ric Fenton of the National Mining Association (NMA), an organization which represents 70 percent of the nation's coal mining companies and 60 percent of those involved in hard-rock mineral production. A heavily built man with an accent that fixed him firmly to West Virginia, he agreed that what had happened at Summitville was unacceptable. "Just so you know," he drawled, "we don't tolerate that kind of behavior among members. He was a rogue guy out there. That was wrong."

I asked Fenton why the mining industry was unwilling to pay royalties for hard-rock operations on public lands. It seems I was misinfomed. "We are willing to pay a royalty," he said indignantly. "For eight years now, we've said we're willing to pay a fair royalty on hard-rock minerals, but it needs to be set at a level that doesn't push our members out of business." What the NMA had come up with was a scheme that would involve a royalty on "net proceeds," and it had suggested that the royalty could be set at 5 percent.

The Mineral Policy Center has long advocated that hard-rock mining companies should pay a net smelter royalty of 8 percent— in other words, 8 percent of the value of the minerals, less the cost of smelting. This, said Boulanger, would bring hard-rock mining companies in line with coal companies, who pay an 8 percent royalty. However, the mining industry had persistently resisted this, arguing that the extra tax burden would make many operations uneconomical. "The coal industry used to say that if it was subject to a royalty, they'd be driven out of business," said Boulanger. "But they've done fine, even with the 8 percent royalty." According to the Mineral Policy Center, an 8 percent royalty on hard-rock minerals would have raised US$230 billion for the government during the past century.

Fenton had heard this argument many times before, and he produced a fact sheet to illustrate why hard-rock mining operators should not be subject to the sort of royalty arrangements that applied to oil, gas, and coal. Coal and crude oil and gas are salable products as soon as they come out of the ground. In contrast, a number of costly processes are necessary before hard-rock minerals can be turned into something of value. This adds significantly to the costs of production. This is why the NMA wants a net proceeds royalty.

The Mineral Policy Center recognizes that more processing is involved in the hard-rock mining industry, but it maintains that the public would be paying mining's business costs if the net proceeds royalty were used. Furthermore, the allowable deductions that Senators Craig and Murkowski introduced in a 1997 bill (had it passed, it would have introduced a net proceeds royalty) included a whole raft of vague expenses such as "miscellaneous costs," as well as transport costs, exploration costs, mill and mill operating costs, depreciation, and so forth. As far as the Mineral Policy Center is concerned, there is a lot of wriggle room here, and the suspicion is that if the net proceeds royalty ever becomes law, then mining companies will load their costs into the calculation in such a way as to ensure that they pay a minimal amount of royalties.

According to Fenton, the 8 percent net smelter royalty would push many mining operations into the red, and companies would be forced to relocate abroad. From an environmental point of view, he thought this would be disastrous as "99.9 percent of the time we handle environmental matters well in this country," and "That's not true for a lot of other places." Besides leading to a significant decline in mining activity, the net smelter royalty would also cause a significant loss of jobs in areas where entire communities depend upon mining. "Everyone talks about the great American economy," he said, "but what builds the great American economy is people who produce things."

I can understand some of the fears that Fenton expressed about the Mineral Policy Center's approach. Perhaps an 8 percent net smelter royalty would make some operations unviable; but, then, those operations with the smallest margins are often the ones that make the least sense environmentally, as well as economically. Perhaps some companies would end up with their principal operations elsewhere, but American, Canadian and European mining companies are already operating as far afield as Borneo and Botswana, Nigeria and New Guinea. Multinational-bashing is always vogue among environmentalists, but many multinationals are owned by their shareholders and are subject to forms of suasion that do not apply to privately owned local companies. There is, indeed, something to be said for exporting companies who adhere, when abroad, to standards of behavior expected of them at home.

I cannot judge whether the Mineral Policy Center's 8 percent figure is reasonable, but it is clear that hard-rock operations on public land should be subject to a royalty. At present, the Mining Law is handing the mining industry an enormous subsidy. As it happens it probably would be many years before any royalty enriched the Treasury. It is going to take decades, and a great deal of money, to clean up the tens of thousands of abandoned mine sites that litter the country; and both the Mineral Policy Center and the National Mining Association agree that this is how royalties should be used. A Bureau of Land Management (BLM) study in 1998 suggested that the value of minerals taken from public land in 1997 amounted to US$1.2 billion. An 8 percent gross royalty on this would work out at US$96.3 million a year. A 5 percent net proceeds royalty would probably yield one fifth or less of this, although this is just guesswork.

Just as egregious as the give-away sanctioned by the 1872 Mining Law is a rule under the tax code called the percentage depletion allowance (PDA). This allows mining companies to deduct a percentage of their taxable gross income from the taxes that they pay the government. The idea is that this deduction should reflect the declining value of the mineral deposit as it is gradually used up. This might be reasonable if the minerals were owned by the mining companies—in other words, if it had bought them in the first place. But on public land they haven't.

Abolishing the percentage depletion allowance and charging mining companies a royalty for extracting hard-rock minerals from public land might not prevent a few more Summitvilles in the future, but it would at least deter companies from setting up operations that are only marginally profitable and are often environmentally damaging in extreme climatic conditions. Such measures would also provide the Treasury with around US$500 million more a year in taxes—this is roughly what the percentage depletion allowance is costing the government in lost revenues—and with sufficient funds to begin a clean-up of abandoned mine sites that continue to leak toxic chemicals into the rivers and underground watercourses.

So why is the 1872 Mining Law still in place if it has given the public such a raw deal? "That's a simple question to answer," replied Boulanger. "The mining lobby is very powerful and it gives a lot of money to politicians." Certain Western senators did all they could to ensure that the 1872 law was not reformed, and Boulanger immediately mentioned the prominent role played by Senator Ted Stephens, chairman of the Senate Appropriations Committee, and Senator Frank Murkowski, chairman of the Senate, Energy and Natural Resources Committee, both of whom figured in Chapter 1 on the Alaska logging subsidies. One of the most influential supporters of the 1872 law was Larry Craig, a Republican senator from Idaho and a major recipient of political action committee (PAC) money from the mining industry. Over a six-year period, he received US$159,900 of mining money, almost twice as much as the average senator. Between June 1993 and December 1998, mining-related PACs contributed over US$29 million to congressional campaigns, and there were further contributions of US$10.6 million of soft money. The US Public Interest Research Group (PIRG) estimates that US$40 million of contributions helped to secure subsidies worth US$1.2 billion during that period, giving the mining industry contributors a return on their investment of 30 to 1. So far, all attempts to enact reforms of the 1872 Mining Law had been voted down.

* * *

The Mineral Policy Center reckons that there are some 550,000 abandoned mine scars in the United States, and that mines have polluted 12,000 miles of rivers and streams. This claim is vague. A scar might be little more than a scrape in the dirt, and "polluted" could mean a slight increase in acidity. But leaving such quibbles aside, it is clear that the mining industry has often been spectacularly cavalier in its treatment of the environment. If the industry has a bad reputation, as a despoiler of land and a wrecker of communities, it is sometimes deserved, as I discovered when I visited another small community who was suffering from mine-related pollution.

I headed east early one morning, towards San Luis, in southern Colorado. Most of the country I drove through was farmed, and it drew its irrigation water from the Rio Grande and its tributaries, one being the now defiled Alamosa, but there were great swathes of sage brush and scrub forest, too, and one-horse towns with white-painted churches reminiscent of Mexico, who ruled over this area prior to the war of 1848. It took me about an hour to get to San Luis, which nestled in the foothills of the Sierras. I immediately went in search of Bob Green, who worked for the Culebra County Water Conservancy District, and Dan Randolph, the Mineral Policy Center's circuit rider.

Randolph had yet to arrive—he was driving up from New Mexico—but I found Green in his office attending to *La Sierra*, a weekly newspaper that he edits. Produced on a shoe string, it is beautifully written and full of entrancing stories. This week's eight-page edition carried articles on Battle Mountain Gold and its illegal discharges into the Rito Seco creek, on a Civil War re-enactment, and on the importance of Shakespeare to the English language.

A small man with long straggly hair, a lined face the color of a walnut, slanting eyes, and an expressive grin that revealed a jumble of nicotine-stained teeth, Green suggested we go on a tour of the local area while we waited for Randolph. The town itself was unusual in that it had yet to be disfigured by fast-food joints or the advertising paraphernalia that lines many American high streets. There were several cafés and restaurants, most serving Mexican food, and old men in shiny black suits with Homburg hats sat talking in Spanish on the verandahs of wooden buildings. We drove towards the Sierras across La Vega, a large common managed by the community, and wound our way through some fine farmland. Ninety-five percent of the farmers here were Hispanic, explained Green, and they managed their land much as their grandfathers had done. They grazed cattle in lush pastures, and in their small fields they planted beans and corn. "They pride themselves on their farming techniques," said Green. "They all want to prove they can do it in the old-fashioned way without chemicals." Hardly any of the Hispanic farmers used irrigation equipment. Instead, they diverted water— or did, until recently—from watercourses such as the Rito Seco and the San Luis People's Ditch into earthen ditches that crisscrossed their fields.

By the time we got back to San Luis, Dan Randolph, a young man with an earnest demeanor and wry humor, had appeared. Together we set off for Battle Mountain Gold's mine site, which was on open land midway between the town and the Sierras. For six years or so the company had extracted gold using the cyanide vat leach method. However, three years ago the amount of gold in the land being worked over was so small that the company decided to close the mine. Unfortunately, it was already leaking polluted water into the Rito Seco.

The mine presented an extraordinary spectacle. At the center of a large lagoon a pumping device was spraying a great plume of water into the air. The intention was to evaporate as much water as possible, and thus prevent seepage of pollutants into the Rito Seco. Each day half a million gallons, one quarter of the lagoon's water, was pumped airwards. Three-quarters of this evaporated; yet, the water level had only fallen 15 inches during the summer months. The plan was not working, and the company had admitted that the clean-up would take at least 40 years, if not more.

All that separated the lagoon from the Rito Seco was a ridge of land with an unpaved road. "Common sense should tell you that if there's a big hole full of polluted water here," said Randolph, "and a stream over there, then there's going to be seepage. When the state's Division of Minerals and Geology voted to approve the mine, there was one dissenting voice, a local farmer, who said: 'You're not going to be able to control the water.'" He was proved right.

The Rito Seco irrigated the orchards and gardens of some 800 people who lived in San Luis. This year no one had drawn water from the stream; they believed it was too polluted. During earlier times, some of the Rito Seco's water ran into the San Luis People's Ditch, the oldest water right in Colorado; but this year the 20-odd farmers who would normally use the water chose not to do so. They feared it could poison their crops and livestock.

Many local people believed that the regulations imposed by the state were inadequate, and that the Division of Minerals and Geology had acted, here as at Summitville, more as a promoter of the mine than as an independent watchdog. This was a view compellingly put by Jo Gallegos, a farmer whom we called on after we left the mine. "Everything was weighted in their favor," said Gallegos, a handsome character with long black hair, as he puffed a huge Castro-style cigar. Years ago the Mineral Policy Center had hired an expert who predicted that this would happen, but his testimony was brushed aside by Battle Mountain Gold at the committee hearing. The board that oversees the work of the Division of Minerals and Geology was constituted in such a way as to favor the mining industry. Six of its members were miners or ex-miners; only two were farmers, and it was one of those farmers who said that the mine wouldn't be able to control the water. The mining company had persistently initiated operations, and withdrawn scarce water, before it was granted permission, but the state always gave permission retrospectively. In so far as there was a public process, said Gallegos, it was completely one sided and designed to satisfy the mining interests.

Gallegos reckoned that 75 percent of the local community had always opposed the mine; 10 percent were indifferent; and 15 percent were in favor. For six years the mine provided around 30 jobs, but these had all gone now, and the taxes for the local county had also dried up. In return for this, San Luis was left with an abandoned mine that could take a generation or more to clean up, and watercourses that the farmers and the people of the town no longer dared to use.

"Have you heard of environmental justice?" asked Bob Green. It was a concept that sprang out of the belief that big corporations often target their activities on non-Anglo communities, and especially on Indian, Hispanic, and black communities living in remote areas. "This sort of thing would never have happened to candy-assed rich communities like Aspen," he added. Randolph disagreed. He felt that the mining industry went wherever the ore was. Unfortunately, it often happened to be in remote areas where some of the poorest communities lived.

Mining operations are subject to a whole raft of national laws, rather than a single law, and the majority of states have their own laws governing mining activities. However, most of these are inadequate in the opinion of the Mineral Policy Center. It may be, as Dan Randolph said, that hard-rock mining companies tend to go wherever the ore is, but when it comes to coal mining it is clear that companies have sought out—or threatened to seek out—those states where the environmental controls are most lax.

Wally McRae, the cowboy poet who keeps cropping up in this book, spent 30 years trying to get the mining companies operating in his part of Montana to act more responsibly. On various occasions he testified at hearings at the state capital of Helena, and he also traveled at his own expense to testify in Washington, DC, in favor of a reasonable reclamation law. All the while, he said, the mining companies were playing one state off against another. If they could do less reclamation in one state than another, and spend less on pollution controls, that was where they would threaten to go, or where they did go. "The truth is," said McRae, "companies cannot afford to be more responsible than their competitors." This is why it makes sense to have national environmental standards that apply to all states.

"Getting a royalty and giving public land managers discretion to weigh mining against other interests is central to any reforms," suggested Boulanger. "But we also need to establish some baseline national standards for environmental protection." So far, the mining industry had resisted these, arguing that companies already had to abide by a whole raft of laws covering endangered species, waste disposal, clean air standards, and so forth.

One final point needs to be made. It is easy to berate mining companies for polluting rivers, despoiling landscapes, and wrecking the lives of communities who live near mining operations. Such criticisms may, at times, be justified. But we should also remember that mining companies are satisfying a demand, and we as consumers are part of this story. "There is something we need to bear in mind," said Camille Price at Summitville. "We all use metal. We drive tons of the stuff. We wear it. It's in our mouths. We demand it, and that means somebody has to mine it."

In order to get the gold for that small ring on your finger, over 3 tons of dirt and rock had to be moved. It is all very well blaming mining industries for behaving badly, but we are part of the problem, too. When I said something along these lines to two environmental campaigners in Washington, DC, they were horrified. "You sound just like big industry," they chimed in unison. If that's so, then big industry is making a valid point. Consumers, like producers and manufacturers, should be held responsible for their actions.

9

THE PRICE
OF POWER

Just over ten years ago Kitty Myers and her family decided to leave New Mexico and return to Michigan, where her husband Dennis had been raised. They were exchanging a world of bright light and brilliant colors, of sun and adobe, for a damper, more intimate landscape, a landscape that was still wrapped in the gloom of late winter at the time of my visit. They bought a small farm, sight unseen, in the rolling, well-wooded countryside in the north of the state, and since then they have done their best to make a living from the land. It wasn't easy to begin with, and if anything it has got harder since.

When I arrived at the farm, Kitty was feeding her ostriches in a field beyond a jumble of barns and stables. Both they and the cattle looked bedraggled and tetchy, as though they were tired of the snow and mud, of the short days and gray skies. "When we got here, this place was a complete mess," recalled Kitty, a cheerful woman with long brown hair, restrained by an elastic band, and russet-tinged cheeks that suggested an outdoor life. "We found 16 junk cars on the property and we had to re-do the whole house."

Once she had fed the ostriches and stabled a couple of horses, we walked back to the farmhouse, a modest building in the bottom of a small valley. "At the time this was a very peaceful place," continued Kitty. "We'd get a few cherry trucks along this dirt road, but that was about it." Now the Myers found themselves living in the middle of a gas field, and even with the windows and doors shut we could hear the clanking of a horse-head well.

"One day," explained Kitty, "we saw this guy wandering round our fields and Dennis went out to him and said: 'What the hell are you doing?'" The stranger said that he was surveying for a gas well. The Myers knew when they bought the farm that it had what are known as severed rights: they owned the land, but the mineral rights belonged to someone else. They didn't think it mattered. They knew that there had been two wells on this land that exploited the Niagara Reef, but these had dried up many years ago. What they did not know was that a federal tax break was now encouraging a frenetic search for gas in northern Michigan. Over a period of ten years, some 6000 wells were sunk into the Antrim shales, and if you were unlucky enough to be an owner of land with severed rights, there was little or nothing you could do about it.

During the past decade, the lives and property of hundreds of Michigan farmers have been blighted by subsidy-fueled gas development. The Treasury and the American public have also lost out. When the tax credit was introduced, Treasury economists said that it would cost between US$500 million and US$1 billion in lost taxes at most. It is now costing over US$1 billion a year, and by the time the credit expires at the end of 2002 it will have cost the Treasury around US$10 billion.

Kitty Myers had a story to tell that could be told by many others. The surveyors found gas and a company called Savoy acquired the rights to exploit it. Savoy wanted to sink two wells on the 80-acre farm, but after constant pleading from the Myers they agreed to sink only one. All the same, they were determined to put the well where they wanted, in the middle of a hay field, and not, as the Myers requested, at the edge of the property. "The noise caused by the drilling was unbearable," explained Kitty, "and it's been a constant intrusion since."

The men who come to check the well each day park wherever they want; frequently, they leave gates open, even though the fields are grazed by cattle. A neighbor who stood up to a drilling team was constantly threatened, according to Kitty. The drillers defecated on the path leading up to his house and told him: "If you give us more hassle we'll see there's a processing facility in your backyard." Each unit (a unit consists of 10 to 30 wells) has a processing facility and a compressor run by a huge motor. The one that serves the Myers' well is over 1 mile away; but even at that distance it is loud enough to disturb them at night. During the summer they sleep with the windows closed and the fans on to drown out the noise.

The Myers lost over 1 acre of land, for which they were given just US$700 in compensation, and they have spent considerable sums of money insulating the house in the hope of getting a better night's sleep. They have even built a barn between the farmhouse and the gas well to shield them from it. It was obvious that they were not wealthy folk. When I visited their farm, I noticed that one of their tractors was over 40 years old, and a seed drill was of a vintage that would attract museum curators.

As I left the Myers' farm, and headed back to my hotel on the shores of Lake Michigan, I reflected on the injustice meted out to the Myers and many others like them. Wendell Berry has expressed the vanquishing of the decent by the powerful far better than I can. "If there is any law that has been constantly operative in American history," he wrote in *The Unsettling of America*, "it is that the members of any *established* people or group or community sooner or later become 'redskins'—that is, they become the designated victims of an utterly ruthless, officially sanctioned, and subsidized exploitation."

* * *

If you are looking for a headline to describe what has happened to the Myers, and to others in a similar situation, you might come up with something along the lines of: "Industry uses tax dodge to exploit land and people." You might, on the other hand, plump for: "Tax credits help industry provide clean source of energy." Neither statement is any more or less true than the other, which makes the story of gas exploration in northern Michigan especially intriguing. The industry has undoubtedly had a considerable impact on the lives of landowners, especially those with severed rights, and the companies involved in gas exploitation have done their best to pay owners of mineral rights as little as they can get away with. There has also been an environmental price to pay. But of all the fossil fuels, gas is by far the cleanest. If governments are going to provide subsidies for energy producers, it makes sense to favor the gas industry. Of course, whether there should be any subsidies at all for energy producers is entirely another matter.

Energy subsidies come in various shapes and guises. In the Western world most are designed to provide direct benefits to energy producers and to stimulate production. In the developing world, and in the old socialist economies, a deliberate policy of under-pricing provides consumers with a subsidy. In the United States, tax breaks such as the percentage depletion allowance awarded to oil, gas, and mining companies, and the tax credit that applies to gas developments in Michigan, are highly significant. There are also direct subsidies whose purpose is to help the industry research and develop a range of energy sources. Since 1948, the US Department of Energy (DoE) has spent over US$110 billion on research and development (R&D), and over four-fifths of this has subsidized the nuclear and fossil-fuel sectors. The former has been a controversial and costly failure. The latter has provided the bulk of the country's energy; but the extraction and distribution of fossil fuels have caused many environmental problems. In the 50-year period up to 1998, the nuclear industry received US$66 billion in subsidies and fossil-fuel industries received US$26 billion. During the same period, the government spent US$8 billion on energy efficiency measures and US$12 billion on R&D programs for renewable sources of energy.

When I first began to cast about in search of an energy subsidy to investigate, I thought I would be spoilt for choice. Should I head for one of the many research establishments involved in the DoE's Clean Coal Technology Program, a contradiction in terms if ever there was one, coal of all types being one of the dirtier fossil fuels? Or should I make my way south to Tennessee, where the US government was paying British Nuclear Fuels Ltd. some US\$284 million to turn radioactive waste, whose production had already been heavily subsidized by the taxpayer, into frying pans, knives, forks, and other implements essential to a well-stocked kitchen? On the other hand, suggested campaigners in Washington, DC, I might like to investigate the way in which nuclear waste was being reprocessed in South Carolina at a cost to the taxpayer of US\$359 million a year. Then there were the oil R&D programs that were proving such a boon to corporations such as Texaco and Chevron. In the view of the Green Scissors campaigners, all of these subsidies were perverse. They were bad both for the taxpayer and for the environment.

But what would I see if I visited these sites? Not much. I could peer through the razor wire and interview scientists and bureaucrats; but establishing a direct link between the subsidy and specific environmental damage would be difficult. Coal-fired power stations emit large quantities of carbon dioxide, but the nature of air pollution means that gases are widely dispersed, rather than concentrated on one area and one community. What I was looking for, I told FoE's economics campaign director, Gawain Kripke, was a subsidy whose impact on both people and the environment was obvious. He suggested that I get in touch with the Michigan Land Use Institute. In 1997 the institute had collaborated with Taxpayers for Common Sense, FoE and 20-odd others to produce *Green Scissors Michigan*. Modeled on the national campaign, the Michigan report highlighted subsidies that were costing taxpayers some US\$2.8 billion, and which were wrecking the environment. One of these was the cumbersomely termed Section 29 Unconventional Fuels Tax Credit for Natural Gas Development.

*   *   *

The men who run Michigan's oil and gas industry physically rankle at the mention of Keith Schneider, while those who have suffered at their hands shake their heads in wonder when they recall the battles that Schneider's organization, the Michigan Land Use Institute, has won on their behalf. You only have to spend a few minutes with Schneider to realize that he is clever, pugnacious, and someone you'd rather have as an ally than as an opponent. A chunkily built man with dark hair, pronounced features, and a sharp wit, Schneider had spent much of his life—he was now in his mid-forties—working as an environmental journalist for *The New York Times*. During the early 1990s, tiring of life in Washington, DC, he and his wife Florence decided to move to the north-west coast of Michigan's lower peninsula. "It had everything we wanted," he said as we ate lunch in the pretty little lakeside town of Frankfort. Beautiful countryside, clean air, and fine food all sprang to mind straight away. "Florence and I come here most weekends," he added as we savored quail stuffed with duck's liver.

Over lunch, and during the course of the afternoon, Schneider told me how he became involved in the war against gas subsidies. It was a long and convoluted tale, and it gradually unfolded against the scenic backdrop of his new home. After lunch we drove down to the shores of Lake Michigan. The landscape had a washed-out look, with mist rising off lozenges of melting snow and livestock hock-deep in mud, but the somber weather could do nothing to mask the beauties of the coastline. Golden sands backed by heaving dunes stretched south as far as the eye could see. On the way back to his office in Benzonia, we passed a large rambling hotel by the lake. "That's where we spend our summer vacation," explained Schneider. He and Florence lived just 11 miles away, but they loved the area so much that they vacationed here, too.

In 1993 Schneider bought a house and 90 acres of land. He checked to make sure no highways or power lines were planned.

"In fact, I checked just about everything," he reflected, "but I didn't even think to check for oil and gas development." So they left Washington, DC, and moved in. Schneider continued to travel for *The New York Times*, flying in and out of Traverse City with his laptop and a keen nose for a story. Then, one day in February 1994, a landsman employed by the gas industry knocked on his door. "He said everyone was going to get rich," recalled Schneider. "I asked him in and just let him talk for two hours. That knock on the door changed my life."

There were ten landsmen working their way through the area, signing up whomever they could for the rights to exploit gas. "They came through like a squadron of invaders," said Schneider. He took time off work and traveled inland to Montmorency County, where drilling and gas exploitation were already well underway, and he arranged meetings with landowners and others concerned about gas development on his home patch. "The gas companies had already turned a beautiful area of wilderness in Montmorency County into an industrial zone," explained Schneider, "and I was not going to let that happen here." To cut a convoluted story short, a coalition of different organizations began to air their concerns. The debate between landowners and environmentalists, on the one hand, and the oil and gas industry, on the other, became increasingly acrimonious, and the local townships set up a task force to negotiate a workable land-use plan that would satisfy all sides. "'You can get the gas without ripping off people and wrecking the environment,' we said," recalled Schneider, "but the industry just said: 'Get lost!'" And that was when the battle really began. In May 1995 Schneider took the lease of the old one-room schoolhouse in Benzonia—citizen action had recently prevented McDonald's from turning it into a hamburger joint—and founded the Michigan Land Use Institute. It was only then, when the institute began working on the gas issue, that Schneider discovered that a federal tax credit was the catalyst for developing the Antrim shales.

If you want to stretch a point, you could blame the Arabs. In 1973 Saudi Arabia and other major oil-producing countries reduced the supply of oil. Prices quadrupled and Western nations began to look carefully at reducing imports and increasing home supply. The oil embargo, and a shortage of natural gas supplies during the late 1970s, prompted Congress to pass the Crude Oil Windfall Profits Tax Act in 1980. One of its provisions, known as Section 29, was designed to encourage the exploitation of hitherto hard-to-tap natural gas reserves in the tight sands of New Mexico, the coal-bed shales of Alabama, and the Antrim shales of Michigan.

The industry had actually known about the Antrim shale gas since the first wells were sunk in the 1920s. To get to the Niagara Reef, 1 mile below the Earth's surface, wells had to be sunk through the Antrim shales. Gas would seep out of the shales and up to the surface, but only for a while. The shales were so dense that they had to be broken up to yield gas as a continuous flow. Before the introduction of the tax credit, exploiting the Antrim shales was prohibitively expensive and the industry had yet to develop a technology with which to do it effectively. The tax credit changed all that.

This is how it worked. Providing that the wells were sunk before 1992, investors were allowed to knock US 52 cents off their tax bill for every 1000 cubic feet of gas taken from the shales. Indexed to inflation, the credit had a value, at the time of my visit, of over US$1 per 1000 cubic feet. With the average price of gas fluctuating around US$2.50 per 1000 cubic feet, this meant that the credit added around one third to its value.

Scroll back to the 1980s. Let us say that you are a Hollywood film mogul or a Washington attorney or a porn queen. Your tax bill is enormous; you want to reduce it. Your tax adviser suggests—or suggested, prior to 1992—that you invest US$250,000 in the Michigan gas industry. This is roughly what it costs to drill a well. You provide the money to one of several small companies (the major companies, with the exception of Shell, have stayed out of the Michigan gas business) and it sinks the well, links the well to a processing facility, removes the carbon dioxide, and sells it to the utilities. If it is an average sort of a well, it will yield 100,000 cubic feet a day, or around 36 million cubic feet a year. That works out at around 36,000 units of 1000 cubic feet, each worth just over US$1 in tax credit. Each and every year until 2002, when the credit expires, you knock between US$36,000 and US$40,000 off your tax bill. In six years you have paid off your investment. The rest is profit, and to this you must add the annual royalty checks. It is a good deal—a hell of a good deal. Of course, the Treasury is losing out, but then your money has helped the industry to get gas out of the ground.

If I wanted to hear the industry's point of view, suggested Schneider, I should call on Greg Fogle, the president of O.I.L. Energy Corp., whose offices were in Traverse City. He described Fogle as rabidly right wing. When I visited him, Fogle was affable and forthcoming, and clearing his desk in readiness for a vacation in Florida. Had it not been for the tax credit, said Fogle, the oil and gas industry in Michigan would have died a death during the late 1980s as the Niagara Reef began to dry up. In those days the industry employed 12,000 to 15,000 people; now, it provided jobs for 7000 to 8000, and it would have been many fewer had the industry failed to tap the Antrim shales. The tax credit attracted the investment that was required for companies such as his to develop the technology to extract the gas. "Few tax credits have worked as well as Section 29," suggested Fogle, who at one time ran 600 wells. It had achieved exactly what Congress had intended. Without it, there would have been no gas industry in Michigan. And without it, Montmorency County, where I headed once I left Fogle, would still be a rural backwater.

Atlanta, Michigan, sits at the heart of Montmorency County. It is a one stoplight town. In the hunting season it is overrun with men in camouflage gear; but now, in late winter, it was struggling slowly out of hibernation and was about as lively as the stuffed elk on display in a glass cabinet at the main crossroads.

Schneider had told me to get hold of Tom Edison and I found him waiting for me in the offices of the Montmorency County Conservation District. A stocky individual with a silver beard, he introduced me to his colleague, Donna Hardin, a sallow-complexioned woman with dark brown hair. "I have to warn you," said Edison in a deep, husky voice, "that I've blocked out of my memory a lot of what happened here. It was one of the saddest experiences of my life when the oil and gas industry came." He was a little shaky on dates—Hardin helped him with these—but he could readily recall the emotions of those troubled times. "The development came like a wave," he remembered. "It all started in Otsego County, to the west of here, and the gas industry said it'd stop there. It didn't, though. When it came to Montmorency County the impact was right here, out in the open for everyone to see."

Noise was a major concern, both then and now. There were several hundred injection wells in the county, each serving between 10 and 30 individual wells, and each with up to three large engines running continuously. "One day a retired policeman came into this office," recalled Edison. "He'd worked in River Rouge, a rough community near Detroit, and he'd sunk his savings into buying a retirement home up here. The year after he bought it, they put in an injection well next to his house. He came to this office in tears. 'Listen,' he said, 'I've shot people before. I don't know who to shoot now.' He was frantic."

According to Edison, the gas development had had a devastating impact on the environment. "Once you could walk 10 miles without crossing a road," he said. "Now, if you look at an aerial photo, it looks as though someone's sneezed all over it." There were now more than 700 miles of pipeline in the county, and hundreds of miles of footpaths had been turned into two-lane tracks. In the early days of the development, the drilling crews bulldozed through rivers and streams to lay pipelines, and there was a huge amount of erosion caused by all the work. It was true, said Edison, that the amount of land taken up by the wells was not that great. However, each well was served by a network of roads and pipelines, and until the early 1990s, when eligibility for the credit expired, the industry sank wells at a density of one every 40 acres. As soon as eligibility for the tax credit expired, spacing between the wells in many areas quadrupled from 40 to 160 acres.

For Schneider, this was proof that the gas companies were "mining the tax credit." The quicker they got the gas of the ground, the more they got in credits. In 1995 a court order forced the regulatory body, the Department of Natural Resources (DNR), to increase the spacing of wells from one every 40 acres to one every 80 acres. Significantly, Shell was already sinking just one well every 160 acres, which was what environmentalists were calling for. If a company like Shell could do this, asked Schneider, why couldn't others?

In Edison's view, the gas industry had behaved atrociously in its dealings with local people. "This is a very arrogant group of people," he said with unconcealed anger. In Montmorency County, most of the land and the mineral rights belonged to the state, and initially the gas companies made no attempt to consult the local people. Edison felt that the DNR and the Department of Environmental Quality, which was carved out of the DNR during the mid-1990s, had failed to regulate the industry in a way that he and others had a right to expect. "They are very much in somebody's pocket," he said at one point, and it was clear who he thought that particular somebody was.

Many of those I met were especially scathing about the landsmen who were sent out to do deals with private owners of mineral rights on behalf of the gas companies. "They were clean-cut, articulate, and bright, and they would lie to you without... ." Lost for words, Donna Hardin let out a long whistle. "Jesus, they would give you the creeps!" On the first visit to a mineral rights owner, they would be charming and solicitous; but if the owners refused to sign a contract, or argued for a better deal, the landsmen became progress-ively more unpleasant. Soon after it was founded, the Michigan Land Use Institute developed what Hans Voss, Schneider's deputy, described as a "how-to leasing package" for landowners. "I had farmers ringing me all the time, asking how they should deal with the landsmen," explained Voss. "I'd say: 'You're in a great position. Don't sign their piece of paper. Here's how you can double what they're offering you.' This was a very significant shift. It made people realize that it was their land, that they had control over it."

*   *   *

The year 1993 was a tough one for the gas industry in Michigan. New wells were no longer eligible for the Section 29 tax credit, and the price of gas was falling. Before December 1992, there had been a rush to sink as many wells as possible in order to get the credit. In 1991, some 1200 permits were issued; in 1992, there were over 2000. The following year the number of permits issued by the DNR fell to under 650. But then a funny thing happened. By 1995, the number of permits was back up to over 1500 and the industry was booming again. Why? Because behind closed doors in the state capital of Lansing the industry had struck what Schneider called a sweetheart deal with the state government. The DNR agreed that the industry could write off what were described as post-production costs, and the royalties and taxes owed by the gas companies to the state would be calculated after these costs had been deducted. "These costs included just about everything you could think of," said Schneider. "Cleaning the bathroom. Installing and using phones. Plowing snow away." According to Schneider, the state's officials had told the industry negotiators that the post-production costs only applied to state-owned minerals, and they were warned not to introduce them in their dealings with private owners of mineral rights. They ignored the warning and were found out in a big way.

"Some time in 1995," recalled Hans Voss, "we had farmers ringing up this office, complaining that their royalty checks were all of a sudden way down. The wells were still pumping away, and gas prices hadn't fallen enough to explain the drop in royalties." Voss and

Schneider began to dig around and they soon learnt of the deal made between the state and the industry. Without so much as a word of warning, the industry had now begun to apply post-production costs to private mineral rights owners, too. Let's say the value of the gas, prior to the deal, was US$50,000. The mineral rights owner would get one eighth in royalties, or US$6250. The gas companies might knock off, for example, US$20,000 in post-production costs. That meant that the mineral rights owner got one eighth of US$30,000, or US$3750, and the companies paid taxes on US$30,000, not US$50,000. As a result, the state lost out, too.

I wanted to hear, first hand, how landowners had suffered, and Voss pointed me in the direction of Ed Lennington, whose experience was similar to that of many other private mineral rights owners. Lennington lived on a farm near Mancelona, some way north of Traverse City, but he didn't farm in a conventional sense. He ran a dental workshop from his double-wide trailer and the only things he grew with a profit in mind were ginseng and golden seal, whose seeds he scattered in the woods and which he hoped eventually to harvest and sell. When I arrived, Lennington was pottering about in the kitchen, which had an old-fashioned juke-box near the cooker. A powerfully built man with silver hair and close-set brown eyes, he suggested we drive downtown to a small diner.

He was not a greedy man, he explained as he ordered a burger, but the size of his royalty checks mattered, not least because his wife was terminally ill and he needed every dollar he could get to look after her. Initially, all went well. He received over US$4000 in royalties the first year, and the company said it could only go one way after that: up. For a while, that is exactly what happened; but without warning he received a statement which showed that around US$500 had been knocked off the US$1800 he was owed for that month. "I called them up," he recalled, "and they said: 'That's our expenses.' I said: 'Hold on, I have sold you gas. Your expenses are your business, not mine.'"

That was in 1994, and since then he had received ten statements in which he owed the gas company money, rather than the other way round. Lennington told his local political representative: "We need to get together, 'cos if this doesn't stop, you sons of bitches will never get re-elected." Lennington also wrote to Governor John Engler about the post-production costs, and Engler wrote back to say that he would be hearing from a Rodney Stokes. "This Mr. Stokes wrote back and what he said was: 'It's a very complicated process, and it would take too long to explain it to you, so you'll just have to trust us.' I know we're stupid dirt farmers. . . but for Christ's sake!"

Lennington and hundreds of others felt that the gas industry was taking them for a costly ride. There was even talk of blowing up the wells. "Not a good idea," as Lennington reflected. "That would have led to houses getting blown up, too." More sensibly, the Michigan Land Use Institute arranged meetings that were to provide mineral rights owners with the opportunity to question the industry, state regulators, and politicians. The post-production costs were now firmly on the political agenda; and according to the state's own calculations, the deal, negotiated in private by Engler's Republican administration and the industry, was costing the state at least US$4 million a year in lost revenues. "In my view," said Schneider, who put the figure closer to US$8 million, "there was serious and transparent political favoritism. The Engler administration has a long history of responding to its most generous political donors by opening the state's treasury for subsidies, as in this case, or by transferring large parcels of public land to private interests." He added that between 1990 and 1998, Governor Engler had received upward of US$380,000 in contributions from the Michigan oil and gas industry— not, in the overall scheme of things, a colossal sum of money, but enough to ensure that the state acted towards the industry with considerable sympathy.

Agitation by Lennington, Schneider, and others paid off. Once the scale of the post-production write-off was exposed, the director of DNR reduced the allowable post-production costs on state leases by two-thirds.

Lennington used an analogy to explain the inequities of the post-production system. "You're a shoemaker," he said, "and I need a pair of shoes. I like the ones you are selling for US$20, and I ask you how much they cost to make. You tell me that your expenses amounted to US$12. I say: 'OK, I'll give you US$8. You can bear your own expenses.' That's exactly what the gas companies are doing to us." Later, when I returned to Benzonia, I repeated this to Schneider. "That's exactly right," he said. "That's a very good analogy." Later still, in the offices of the Michigan Oil and Gas Association (MOGA), I repeated the comment to its chief executive officer, Frank Mortl. He was unimpressed.

Mortl's office, high up in a Lansing skyscraper, commanded fine views of one of America's dullest cities. I could sense, as soon as we shook hands, that he was feeling edgy. "Are you with McCain?" he inquired, peering at me through heavy-rimmed spectacles. "Are you writing this thing for McCain?" By chance, I had come to Michigan a week after the Republican primaries, during which Senator John McCain of Arizona had said he wanted to eliminate tax breaks for big business. Besides declaring war on corporate welfare, McCain had also vowed that if he were ever elected president he would reform campaign finance and sever the financial links between big business and politicians. His views were not designed to please men such as Frank Mortl, or the two individuals he called up to participate in our conference-call discussion. Mike Miller, the chairman of MOGA and president of Miller Energy Inc., sounded like the sort of person who wouldn't countenance bluster or small talk. He was a serious player in the oil and gas business, and afterwards Mortl told me that his father, C. John Miller, had had almost as much influence on Michigan's industrial development as Henry Ford. At Miller's side, and guardian of his company's facts and figures, was his finance officer.

The Section 29 credit had brought tremendous benefits to Michigan, according to Miller. "It was a great deal for industry, consumers, and the environment. Gas is the cleanest fossil fuel we have and the credit helped us to tap the Antrim shales," he said. Nor was there anything untoward about the industry setting its post-production costs against the value of the gas before paying royalties and taxes. What I needed to understand, said the three men, was the history of gas delivery. During the 1980s, the utilities that bought the gas came right up to the well head to get it. "It was a very simple, very easy world," chipped in Miller's finance officer. "The gas had a value at the well head. Now 99.9 percent has no value at the well head, because the utilities are not buying it at the well head." Greg Fogle had told me a similar story. Not only did the gas utilities tell companies like his that they were going to have to pipe the gas, at their expense, to a place which suited the utilities, they were also told that they had to remove the carbon dioxide from the gas, which previously they hadn't. In short, for the gas to be a salable product, the gas companies had to do more and spend more. It was therefore only fair that all the other beneficiaries, including the private mineral rights owners, should share the costs of delivery. "The way it works out," added Mortl, "is that I'm paying seven-eighths of the post-production costs and the owners, who get an eighth royalty, are paying an eighth."

I told him about Ed Lennington's grievances about the post-production costs. "There's maybe a misunderstanding of a complex issue there," he replied. "This guy from Mancelona, maybe he got more information than he should have got off of the computer." I also mentioned that the environmentalists were greatly exercised by the fact that gas companies had charged for all manner of things, from snow plowing to phone bills. Wasn't this going a bit too far? "No," he replied, "the phone is hooked to a circuit and in an emergency it would ring someone automatically if the flow of gas stopped. It's a legitimate business cost."

So what about Schneider's contention that the gas industry, spurred on by the federal tax credit, and later by the post-production cost savings, had sunk far more wells than were required to get the gas out of the ground? "The tax credit is paid on the amount of gas produced, not on the number of wells," answered Miller abruptly. During the early days, the industry spaced wells out every 40 acres, because the technology did not exist to enable them to space them out at greater distances. But now, with horizontal drilling and other technological advances, they could space them out more; and in the view of these men, the environmentalists had greatly exaggerated the impact of the wells. True, said Mortl, noise could be a problem, but nowadays the gas companies did their best to buffer their machines, site the wells away from human habitations, and erect boards to stop the sound of compressors from causing too much disturbance.

I told Mortl that I had heard plenty of criticism about the landsmen. "Yeah," he sighed wearily, "I've heard that too, but I've never managed to pin it down. I'm not telling you everybody is pristine and pure. You may find a landsman that was gruff, or one of the follow-up crew, but I haven't ever had anybody give me their names." He said he had a 24-hour toll-free number that people could ring to make complaints. If individuals within the industry were behaving badly, his organization would soon sort them out.

*   *   *

So where does all of this leave us? There seems to be precious little common ground between the Michigan Land Use Institute—"the so-called, self-declared Michigan Land Use Institute," as Mortl called them contemptuously—and the oil and gas industry, most of whose leaders, in the view of Schneider, were rabid right-wingers.

Let us take, firstly, the environment. We are not, as it happens, talking about one of the great landscapes of the United States, or even one of the great ecological systems. A century ago, the whole of northern Michigan was plundered for timber and I did not see one patch of primary forest. Rather, I drove through oceans of secondary growth, of pine and spruce and cedar. Having said that, the gas industry has transformed large areas of wilderness into a road-hatched, well-spattered, not particularly wild place. Schneider may be using hyperbole when he describes Montmorency County as an industrial zone, but he is right to suggest that the gas industry, spurred on by the federal tax credit, sank far more wells, perhaps three times more, than were required to tap the gas. With fewer wells, it would have taken longer to get the gas out; but why the rush? There may have been a shortage of natural gas during the late 1970s, but there was a glut of the stuff in the early 1990s. As Miller succinctly puts it: "In the early days, people said they'd never find oil and gas in Oklahoma or Louisiana. That proved to be totally false. We're always finding new deposits." In fact, it was the availability of cheaper gas from other parts of the United States that enabled the utilities to impose greater demands on Michigan's gas companies, and which led to the introduction of post-production costs. Admittedly, it looks as though demand is set to outstrip supply over the coming years; but this is still no excuse for causing unnecessary environmental damage. So, yes, the gas industry has had an adverse impact upon the environment and the tax credit had much to do with this.

It is people, however, who have come off worse. The industry has brought absolutely no benefits to people such as the Myers, who have severed rights, or to the retired policeman from River Rouge, who was eventually driven out by the noise of the compressors. I have no idea why so few people have used Mortl's toll-free number to complain about harassment by landsmen and others. This may be due to fear, or perhaps few people believe that the industry will respond to complaints. It is clear that the industry and its employees frequently acted in an appalling manner; but there have been winners as well as losers. A considerable number of private mineral rights owners have benefited from gas exploitation, and the royalty checks have brought a degree of wealth to farmers struggling to survive. In these times of falling commodity prices, every cent helps. The state, too, has benefited from the oil and gas industry. Over the years, royalties from the industry have provided US$550 million to the Michigan Natural Resources Trust Fund, which has been used to buy up and improve around 135,000 acres of land that are now open to the public. The industry also pays severance tax to the Treasury and a surveillance fee, which covers the cost of running the Department of Environmental Quality, which is supposed to regulate the industry.

However, few if any benefits have trickled back to the counties and townships. Montmorency County has been transformed by gas development over the past two decades. There are now more pipelines than county roads, and there are over 2000 wells, one for every five people living in the county. This is one of the poorest areas in the state and desperately short of funds with which to run its schools and other public services. The gas industry and the investors have made a fortune from the county, said Tom Edison, and in return the local community received nothing. In fact, the industry showed what it thought of people such as Edison by naming one cluster of wells the Light Bulb unit. The joke may have amused the industry, but Edison thought it crass.

As far as post-production costs are concerned, I can see no good reason why the industry should not be able to set some of its legitimate costs against tax and royalties, just as freelance writers are able to set travel, typing, and other expenses against tax payments. Nor do I think it unreasonable to expect mineral rights owners to share a portion of the costs of getting gas from the well head to the utilities—and the portion should be set in line with the royalties. However, the gas industry was disgracefully greedy, and this was confirmed by the DNR ruling that reduced its allowable costs by two-thirds. By piling on virtually every expense they could think of, the gas companies were not only fiddling the state out of royalties and taxes, but robbing private mineral rights owners of their rightful share of royalties.

*   *   *

Worldwide, energy subsidies probably amount to some US$250 billion a year, and one of the starkest examples of how ridiculous some of these subsidies are comes from that outwardly most rational of nations, Germany. To foster home production and keep the coal industry in business, the government provided US$42 in subsidies for each ton of coal produced during the early 1980s. By 1996, the cost of this protectionist policy had risen to US$153 dollars a ton, and the subsidy bill had risen from US$4 billion to US$7 billion a year. Then, in 1997, the government decided to cut producer support and halve the number of jobs in the industry by 2005. However, rather than reduce the bill, early retirement schemes and the like pushed the cost of supporting the coal industry up to US$12 billion a year. What is more, German taxpayers have actually been shafted twice. Firstly, they have propped up an uneconomic industry. Secondly, they have been called upon to subsidize the development of wind power as an alternative source of energy. And they have had to subsidize it heavily in order that it can compete in the market with subsidized coal.

By providing the bulk of subsidies to nuclear and fossil-fuel programs, governments have put the renewable energy sector at a competitive disadvantage in precisely the same way that sugar subsidies in Europe and the United States have put unsubsidized Caribbean cane producers at a disadvantage. Furthermore, conventional fuels in the United States are unrealistically cheap, and this, too, discriminates against alternative fuels. "There's a need to develop alternative energy sources," suggested Mike Miller in Lansing, "but as long as we hold the prices [of conventional fuels] artificially low, we're not going to." And here speaks one of the big players in the oil business.

In the United States, the bulk of energy subsidies have gone to the nuclear and fossil-fuels sectors, and a good chunk of this has been spent on R&D. The DoE argues that because R&D programs are so costly and high risk, no single company could afford to do the sort of research that it (as well as the taxpayer) is bankrolling.

However, this argument does not impress analysts at the Massachusetts Institute of Technology. In a recent report they suggest that the "experiences of the 1970s and 1980s taught us that if a technology is commercially viable, then government support is not needed, and if a technology is not commercially viable, no amount of government support will make it so."

There are compelling fiscal and environmental arguments for scrapping all subsidies that go to the nuclear and fossil-fuel sectors. As far as the former is concerned, subsidies have been a monumental waste of money. As far as the latter are concerned, they have led to artificially low prices, and encouraged higher levels of consumption, and therefore pollution, than would have occurred in an unsubsidized world.

But should governments subsidize the development of alternative sources of energy? Most environmentalists maintain that the environmental benefits of alternative energy sources generally outweigh the cost of the subsidy; and they argue that if alternative sources are to compete in a market dominated, at present, by big plants, they will need some help. Unfortunately, many subsidies for alternative energy research have yielded little of value. For example, the United States spent US$1.4 billion over two decades developing solar energy mirrors in the Mojave Desert. Thus far, nothing of the slightest commercial sense has come out of the project. Over the past 20 years, DoE subsidies to promote renewable energy have exceeded US$11 billion, and yet these alternatives have captured just 2 percent of the electricity market.

I suggest that the best thing we can do is to abolish all subsidies for fossil and nuclear fuels, and introduce measures such as carbon taxes, which would ensure that the price we pay for fossil fuels reflects the external costs that they impose upon the environment and our health. This would go some way towards helping renewable forms of energy to become more competitive. It seems, however, that the Bush administration has other plans. If the House energy bill, HR4, is signed into law—it was under consideration at the time of going to press—subsidies to the oil and gas, coal, and nuclear power industries could double to some US$60 billion over the next ten years.

During the course of my year in the United States, oil prices rose from US$10 to US$30 a barrel, and gas prices more than doubled. There was much outrage, among both the driving public and in the media. What most people conveniently ignored was that the United States is disgracefully profligate in its use of energy. If the country was more fuel efficient, the demand for energy would lessen, the fuel bill would decrease and energy-related pollution would decline.

Before I set off for Michigan, I spent time with Dan Becker, an energy expert who worked for the Sierra Club in Washington, DC. He was candidly unflattering about his compatriots' use and abuse of energy. Some 4 percent of the world's population is responsible for one quarter of the world's carbon dioxide emissions, most of which derive from the burning of fossil fuels. The United States is consequently the largest single contributor to global warming. Such is the American love for gas-guzzling cars that more carbon dioxide comes out of the tail pipes of sports utility vehicles in the United States each year than is produced by all sources in the United Kingdom.

Americans are not only profligate in their use of energy, they are extraordinarily wasteful, too. According to Becker, Americans use over twice as much energy to achieve a similar standard of living as is enjoyed by the inhabitants of the wealthier European nations. As Becker emphasized, if the United States really tried to conserve energy—by installing double glazing in homes, by insulating commercial buildings, by using energy-efficient light bulbs and appliances, by switching to more fuel-efficient cars—then it could swiftly reduce its consumption by up to 30 percent.

Unfortunately, few people are willing to accept responsibility for their actions when it comes to using natural resources. "For most Americans," remarked Becker dryly, "personal responsibility means single mothers on welfare ought to go to work, not that they themselves should reduce their consumption." By all means, let us berate the energy producers when they carve the tops off West Virginia's mountains to get coal; when oil slicks kill seabirds and damage fisheries; when gas developments turn wildernesses into a pipe-strewn suburbia. But let us remember, too, that as consumers of energy we are responsible, albeit indirectly, for some of these problems.

10

TAKING STOCK

I began writing this book in late summer, after many weeks of travel in the West. I could still feel the warm sun on my back, and smell the tangy scent that wafted around the great sweeps of sage brush and pine through which I had trekked. I could still hear the sounds of the rodeo in Cody, with the star-spangled banner sung country style, and recall the thrill of seeing black bear amble across the forest track in front of me. With me, too, were the memories of time spent in pleasant conversation, with ranchers and wranglers, with waitresses and shopkeepers, with environmentalists and truckers, in small towns and the bars of small towns, where people seemed to have plenty of time for both one another and for strangers such as myself. In short, when I put pen to paper, I felt good about the places I had been to, even if I had, at times, been shocked by the way in which public money was being channeled into needless and destructive activities.

Now, as I sit down to take stock of what I have seen, I find I have none of the jauntiness that I had when I left the West, and I feel much as I did when I arrived in Detroit, once I left the Michigan gas fields. It was late winter, and a bitter wind tugged at a few solitary trees that stood between the shabby motel in which I found myself and a six-lane highway that carved through the eastern suburbs of Detroit. The motel belonged to a landscape of concrete and cars, to a world that valued commerce and consumption above all else. There were vast shopping malls; there were supermarkets the size of Grand Central station; there were enterprises selling cars too numerous to count; there were fast-food restaurants whose food was as tacky as the neon-lit signs that disfigured the skyline. But there was absolutely no sense of community, of people being rooted to place, and only the very poor could be seen moving about on foot.

Early in the evening, I wandered out for a drink, but found no bars and returned to the motel with a half bottle of gin. I switched on the TV, flicked through countless channels whose sole purpose seemed to be to encourage viewers to buy things, then finally settled on a live transmission of Senator John McCain of Arizona campaigning in the Republican primaries. "My friends," he told the Californian audience, "the United States is the greatest nation on earth, and a beacon of hope." The senator soon contradicted himself by saying that the rich were getting richer and the poor were getting poorer. In fact, the United States, like most places, is good in parts and rotten in parts.

If I felt jaundiced now, at the conclusion of my long journey, it had as much to do with what I had seen and heard on my way south to Detroit, as with the dismal surroundings in which I found myself. This is a story worth telling, for it reminds us that subsidies do very real damage to the lives of very real people.

I hadn't intended to look at the hog industry in Michigan, but when I visited the Michigan Land Use Institute I met Patty Cantrell, a tall, good-looking woman with wide-set eyes and a lilting Missouri accent. Cantrell had co-authored a fine critique of the way in which large corporations are squeezing the life out of small hog producers by building what are known as concentrated animal-feed operations. In layman's terms, these are concentration camps for hogs, in which the animals are reared and fattened in oppressive confinement and where the corporations control everything, from the provision of food, often laced with growth promoters, through to the killing and sale of the animals. The concentrated animal-feed operations, or CAFOs for short, had hit the news in a big way the previous fall when Hurricane Floyd battered North Carolina, ripping thousands of hogs from their acrid confinement and sluicing their carcasses into the waterways, along with tides of hog sewage. Since then, explained Cantrell over lunch, many Southern states had begun to tighten their regulations on factory farms, and intensive dairy factories were now moving into Michigan and hog. This trend was welcomed by the business-friendly governor, whose party had recently rescinded laws that allowed townships to impose regulations on things such as industrial livestock operations. There were several subsidies that encouraged this sort of farming, and Cantrell feared that Michigan was about to experience a surge of CAFO expansion.

Cantrell arranged for me to visit neighbors of hog factories on my way south, and I turned off the highway a little way north of the state capital of Lansing. The countryside in North Plains township reminded me of parts of lowland England. The roads were lined with avenues of oak, maple, and walnut, and there was a feeling of permanency about the small farmsteads that were scattered around the fields, most of which were fallow now, neatly plowed in readiness for the spring sowing of corn and beans. Bernadette Fletcher's ranch house was some 2 miles down an unpaved track, and shortly before I arrived I passed a large hog operation, which consisted of four long barns, each of which was served by a tangle of white hoppers. The operation was surrounded by a barbed wire fence and signs that warned against trespass.

We sat on the wooden verandah outside of Bernadette's kitchen, in sight of the small herd of cattle that her husband, an Oldsmobile employee, looked after in his spare time. Sitting here was impossible in the summer, explained Bernadette, a small, vivacious woman with fine brown eyes and a neat figure. "In summer," she said, "the smell is so atrocious, we hardly go out, except to mow the lawn." She had been brought up on a farm, one of 12 children, but she had never experienced a smell like this. The family who owned the nearby hog operation lived on the other side of Lansing, and consequently did not have to suffer the smells. A couple of times a day, someone would come to check on the 4000 fattening hogs to ensure that the feeding and watering systems were working. Although the operation was half a mile from Bernadette's house—some of her relatives lived much closer to the hogs—they could often hear the ghoulish squealing of fighting hogs at night. Their lives had been made such a misery since the operation was set up in 1998 that they had thought of moving. But there was really no question of that, said Bernadette, because this was where they had brought up their younger daughter, who had died five years ago, and this was where the family had built a memorial to her, a stone edifice that was flanked by a tall flagpole on which fluttered the stars and stripes.

Late in the afternoon we drove over to see Jerry and Louise Burns. I imagine they were in their seventies. Both were slim, spry, and agile, and they spoke with passion about the fate that had recently befallen them. We sat in the parlor of their small farmhouse and they told me how a hog operation which had been set up next to their farm in 1996 was ruining their lives. "We thought we had a nice place to live in our old age," said Louise Burns in an anguished voice, "and now this has happened. It's so unfair!" Jerry Burns had been born in this house, and he had fought in North Africa during World War II, fathered eight boys, run a dairy herd, and grown corn, wheat, and soya beans on his 250 acres of land. This had been a good place to live, he said. Their church was just down the road in a small settlement that boasted the oldest bar in Michigan. There was a real sense of community here, and everyone knew each other. But now corporate agribusiness had come into their lives and done as it pleased. Last year, one of the their sons came to visit with his family from California, but they cut short their stay and flew back early. "They could not abide that awful smell," said Louise, shaking her head.

Jerry was not just upset on his own behalf, but for farmers in general. He said that many small hog farmers in Michigan had gone out of business during the past year, and the reason they had gone out of business was because they had been driven out by the big corporations. This is precisely what happened in Patty Cantrell's home state of Missouri, where over one quarter of hog operations with 100 to 500 hogs went out of business in just two years during the mid-1990s.

The corporations' *modus operandi* is simple. They find farmers or landowners, often those who are struggling to survive, and offer them a package deal. "You build the barns and install the equipment needed for a concentrated animal-feed operation, and we will provide the hogs and the feed, and pay you as contract growers." The only things the growers own, apart from the buildings and the mortgage that they will have raised to cover the costs of the buildings, are the hogs that die and the vast oceans of manure, which they must dispose of themselves. When the hogs reach the required weight, the corporations take them to slaughter, often to their own slaughter-houses. This vertical integration of the hog industry may have enabled corporations to produce cheaper meat for the consumer; but the cost had been immense, besides the nuisance and pollution they foist on communities such as those I saw in Michigan.

The corporate hog producers have conspired to kill off the small, independent hog farmers, and they are doing so with the help of both the taxpayer and the political establishment. "If these corporations want to have a vertically integrated hog industry," said Cantrell, "OK, let them do that, but not at our expense." The subsidies they receive vary from low-interest government loans for the construction of hog operations—a small one costs at least US$250,000—to vast sums of taxpayer money spent on research designed to benefit intensive hog producers. CAFOs are, to all intents and purposes, industries, not farms; yet they are subject to pollution laws that were designed for small family farms. CAFOs are also taxed as though they were farms, paying around one third of the standard industrial rate. When you consider that some operations produce more waste than a large city, and that there are plans to build a 2.3 million hog unit in Utah, you begin to realize how ludicrous it is to treat these places as farms.

Approximately nine-tenths of all hogs in the United States are now reared in concentrated animal-feed operations. From the point of view of corporations such as Cargill, Tyson, Continental Grain, and the mischievously misnamed Murphy Family Farms, these vertically integrated enterprises make economic sense. They obviously make sense to the contract growers, too; although some receive a rude shock once they discover what a tough deal it can be. They may find themselves heavily in debt, at the mercy of corporate decision-making, and responsible for the disposal of vast amounts of manure. The corporations would argue that they are providing a service to consumers, and that their produce is cheaper than that supplied by small independent hog farmers. The fact is that consumers often have little or no choice. Visit virtually any supermarket in North America and the only bacon on sale will be the flaccid, tasteless stuff that comes from the country's concentrated animal-feed operations.

The big corporate hog producers are driving small family hog farmers out of business, often at an alarming rate. They employ much less labor per thousand hogs than independent family farms, and they invest far less in the local communities where they operate. While small hog units generally have no difficulty in disposing of manure—on the contrary, the manure is a boon to the land—the CAFOs are causing immense problems in many states. A Premium Standard Farms complex of hog units in Missouri with over 2 million animals generates five times more waste than nearby Kansas City. The nation is pocked with lagoons full of seeping sewage, many of which are leaching their fetid waste into the groundwater.

Big operators such as these are indirectly having a serious impact upon rural communities in other parts of the world, too. The ultimate aim of the corporate hog business is world domination. One day they would like to supply all of us—Americans, Canadians,

British, Peruvians, Japanese, Inuit—with their produce. In the UK, hog producers—or pig farmers, as we call them—are going out of business at an alarming rate, partly because UK supermarkets are buying pork and bacon that is produced more cheaply by their foreign competitors. Many of the latter do not have to adhere to the same strict welfare standards as do the British, nor are they subject to the same strict regulations governing the disposal of waste and the nuisance of smell. Furthermore, many are at a competitive advantage to the British because they are subsidized.

During the past couple of years, Bernadette Fletcher had invited over 20 politicians to visit her and to experience, first hand, the horrors of living beside a corporate hog operation. None had, even though the state capital was less than a 40-minute drive away. She and her neighbors had also written numerous letters to the newspapers. None were printed. They felt that the world was against them. Corporate farming was doing as it pleased, and they were paying the price of powerlessness.

* * *

It is time, now, to reflect on the *dramatis personae*—the fiscal conservatives and environmentalists—who are leading the fight against subsidies and corporate welfare. They are certainly strange bedfellows. Many environmentalists despise the fiscal conservatives, whose free-market, pay-for-what-you-use philosophy seems to them to play into the hands of the big corporations. Most environmentalists see them as being at the cutting edge of capitalism and resource exploitation. For their part, most fiscal conservatives seem to have a pretty low opinion of the environmentalists, whom they regard as intellectually flabby and dangerously keen on big, interfering government—providing, of course, that it interferes in ways which please the environmentalists: that it pays to repopulate the wilderness with wolves, for example, or spends billions of dollars running national parks.

A good fiscal conservative believes in minimalist government and low taxes, and subscribes to a philosophy that asserts that the freedom of the individual is of paramount importance, and that each of us should pay the costs of what we use, whether it is education, health care, forest trails, or toilet paper. Fiscal conservatives are deeply antipathetic to public ownership of any resource, whether this comprises coal, forests, land, or fish stocks. They argue that the market is the most efficient means of allocating resources. Those who deplore this philosophy like to typecast it as right wing. More accurately, it is libertarian. In fact, one of the things that astonished me about the subsidy scandal was that many Republicans, and especially the Western Republican senators, were passionate defenders of a whole range of subsidies, and probably did as much, or more, to keep them in place than did the Democrats. Conventional wisdom would suggest that it would be the other way round, with the right-wing politicians fighting against subsidies, and the left-of-center politicians arguing for their retention.

When it comes to subsidies, fiscal conservatives (as opposed to right-wing politicians) are against them, and if you want to see how they would pare down government and government spending you should consult the *Cato Handbook for Congress*. Many of the perverse subsidies highlighted by the environmentalists' *Green Scissors Report* are here; but the *Cato Handbook* goes much further. It calls, for example, for the abolition of the departments of agriculture, interior, and transportation, and for the privatization of all federally owned land.

It is much harder to give a thumbnail sketch of the environmentalists, or of their philosophy, because they are such a diverse bunch. They range from small single-issue groups with a tiny membership, to organizations such as the Sierra Club, Defenders of Wildlife, the National Wildlife Federation, and WWF, which have millions of members. Some groups, such as Earth First! and Greenpeace, place a heavy emphasis on direct action; others work behind the scenes, trying to bring about change by lobbying politicians. The policies of some groups are shaped by sentiment rather than by science, while others do their best to tease out the truth, rather than pump out propaganda. While some groups concentrate almost entirely on wildlife, others are concerned with issues such as global warming, energy conservation, and the like. Politically, most of the environmentalists whom I encountered liked to think of themselves as "progressive".

As far as subsidies are concerned, the majority of the environmentalists with whom I spent time were not against subsidies *per se*; they were simply opposed to those that cause environmental damage. And in the view of the Cato Institute, they were selective about these, too, eagerly targeting subsidies that benefited Republican constituencies, while ignoring those that flowed into Democratic bank accounts. Be that as it may, most environmentalists saw subsidies as a weapon with which to bludgeon corporations and individuals whose activities they wished to halt. If you can demonstrate that big corporations are not only chopping down public forests, but doing so with taxpayers' money, then so much the better. This is a perfectly reasonable position to take, and a sensible one when it comes to campaigning.

However, simply slashing those subsidies that cause demonstrable harm to the environment would be a job half done. We need to be much more ambitious, and rid the world of all subsidies that influence the way in which we use, and abuse, natural resources. As this tour round North America has shown, subsidies tend to favor the rich and powerful, encourage the overexploitation of resources, and place the subsidized at a competitive advantage against those who are unsubsidized. I should stress here that I am not arguing that the poor, through subsidies, are bankrolling the rich. The poor generally pay little or no tax, and the benefits they receive often exceed what they pay into the Treasury, although they are obviously hurt by protectionist policies that increase the price of foodstuffs and other goods. In a country such as the United States, where 8 percent of the population pays 60 percent of the tax bill, it is the middle classes and the wealthy who are being fleeced in order to provide subsidies to corporations and individuals who could often manage perfectly well without them.

Of course, slashing all subsidies may seem a tall order. If we take into account the power of the special interest groups who benefit most from subsidies, and the wealth that they can use to curry favor with politicians, and then consider the sheer pervasiveness of the subsidy business (worldwide, subsidies probably amount to around 3 percent of the entire gross national product, or ten times more than all expenditure on foreign aid), we could be forgiven for feeling gloomy. Getting rid of subsidies will inevitably be much harder than introducing them; but I suggest that we take heart from some of the subsidy slashing that we have seen over the past decade.

One of the great success stories of subsidy reform comes from New Zealand. By the early 1980s, farmers in this small Pacific nation were as hooked on subsidies as their counterparts still are in Europe, Japan, and the United States. By 1983, farm assistance amounted to around one third of agricultural output, and it was recognized by the farmers themselves, and by their government, that such a state of affairs could not continue. In 1984, massive cuts in subsidies were introduced, and now they amount to a tiny fraction of output. The benefits have been widely felt. The taxpayer is no longer being ripped off. The withdrawal of subsidies has led to a cessation in land clearance, which has been good for the environment. And, most surprisingly of all, the number of farmers has actually risen, rather than declined, as many predicted it would. Operating in a free market, farmers have become more adaptable and more ingenious in order to survive. This is how it should be. All too often, subsidies encourage dependency and sloth.

In the case of the former Soviet bloc countries, subsidies were dramatically cut in response to an economic crisis. These countries simply could not afford them any longer. Between 1990 and 1996, Russian energy subsidies fell by two-thirds, and those of Eastern Europe fell by over a half. This has been good for the environment and for the economies of the subsidy-slashing countries. A similar story can be told for China, who has reduced its coal subsidies by around 50 percent. Its economy is rapidly expanding and there is much less pollution.

There has also been some laudable subsidy slashing in the developing world, too, although progress has been patchy. For example, Brazil abandoned the ranching subsidies that did so much to encourage the deforestation of the Amazon, and many countries have reduced their fertilizer and pesticide subsidies. Several countries and cities have realized that subsidizing water provision not only encourages its wasteful use, but tends to benefit the better-off. As a result, they have reduced these subsidies, and some have decided that the private sector is better at managing water supply than the state.

All of this is heartening, and proves that subsidies can be dramatically cut if the political will is strong enough. We should be encouraged, too, by the fact that subsidies are now firmly on the political agenda. In the United States, the Green Scissors campaign, orchestrated by Friends of the Earth (FoE), Taxpayers for Common Sense, and the US Public Interest Research Group, as well as the work of libertarian groups such as the Cato Institute, has done much to highlight the panoply of perverse subsidies that are doing so much harm to both the economy and the environment. And for some time now, subsidies, and the need to slash them, have preoccupied the WTO, whose task it is to liberalize trade. Many environmentalists are deeply suspicious of the WTO; but they should recognize that it is leading the battle against many of the subsidies that they themselves deplore.

I am instinctively wary of books that are highly prescriptive, which tell readers that this is what must happen, because if it doesn't, the world will go to the dogs. But having dragged the reader this far along the subsidy trail, from Alaska to Florida, from New Mexico to Newfoundland, it would be churlish of me to end the journey by saying: you've seen what subsidies do, now make up your own mind on how to get rid of them. Here, then, are a few pointers about how I believe we should proceed. I start from the premise that, in an ideal world, there should be no subsidies for any activities that involve the use of natural resources. We should ensure that natural resources are priced at a level that reflects their true value, and that the principal of "user pays" applies both to natural resources and to all public works activities, from the construction of dams and irrigation facilities, to the building of roads.

Let us look, firstly, at the issue of pricing resources correctly and take water, a commodity that is in increasingly short supply over much of the world—not least in California and many parts of the American West, where we can expect growing conflict between user groups vying for rights to a dwindling resource. Billions of dollars of subsidies have encouraged the excessive consumption of water and its misallocation. Between 1900 and the mid-1980s, Bureau of Reclamation irrigation projects cost the US taxpayer around US$20 billion at 1986 prices; yet the beneficiaries have so far paid no more than one eighth of the costs. California's Central Valley project cost taxpayers US$4 billion; yet farmers who receive its water often pay as little as a derisory US$10 an acre foot. To put this into perspective, let us recall that the Hispanic farmers whom I met in southern Colorado were paying US$600 an acre foot to a private supplier—and they were still managing to survive. Subsidized water has led to colossal waste, especially in the farming industry, and has encouraged farmers in arid regions to grow water-thirsty crops which in a sane world would only be grown in areas where water and rain are plentiful. If users were to pay the true cost of water—in other words, if prices were set at a level equal to or above the cost of its provision—then this precious commodity would be treated with the respect it deserves. I cannot sum up the issue any better than the *Economist* magazine: "If any subsidies at all are justified, they should go to the poor; rather than money and water flowing to tidy middle-class lawns and acres of alfalfa, both should go to stand pipes in slums."

The principal that users should pay the right price—one that is not distorted by subsidies—for the use of public goods such as water should also be extended to natural resources harvested or extracted from public lands. Why should mining companies be allowed to extract gold and other minerals from public land without paying a fair royalty, as they must on private and state lands? Why should timber on public lands be sold at a price below what timber companies would expect to pay on private land? There is no logic to this. The public is losing out, either because the Treasury is failing to claim revenues that are rightfully theirs, or because taxpayers are directly subsidizing the timber and extraction industries—for example, by paying for the building of forest roads. By failing to charge a fair market price for raw materials, the government is often encouraging companies to exploit resources that would be left untouched if free-market forces were to prevail. Without subsidies, much of Alaska's publicly owned Tongass forest would never have seen or heard a chain saw. Without subsidies, it is doubtful whether Gallactic Resouces would ever have contemplated extracting gold at Summitville, with all the disastrous consequences that this entailed for the environment in southern Colorado.

Does the same apply to livestock grazing, which was the main focus of Chapter 2? Reason would seem to dictate that we take the line suggested by most environmentalists: that graziers on public land should pay the same as graziers on private land. However, as we have seen, public grazing lands are often the poor lands that were unclaimed by homesteaders, and graziers of public land receive few of the benefits accorded to graziers of private land. Comparing the two is difficult, if not impossible, and I would suggest that in this case we adopt a pragmatic approach. The grazing fee system costs more to administer than is received in fees by the federal land managers. If the federal land managers abandoned the system, and allowed permit-holders to graze the land free of charge, it would save taxpayer money. However, there must be a *quid pro quo*. The graziers should abide by strict rules laid down by the federal agencies.

If it is the aim of government is to protect rivers, and riverine vegetation in, for example, the arid South-West from overgrazing, then the authorities should exclude grazing from these areas, or impose stocking rates that ensure that the environment does not suffer.

Nowhere do we need to apply the principal of "user pays" more rigorously than in the realm of road transport. Depending upon whose figures you believe, road users in the United States pay somewhere in the region of 20 percent to 50 percent of the direct costs of providing roads and associated services. There is, therefore, a subsidy ranging from 50 percent to 80 percent. The first thing governments need to do—and in countries such as Japan and France they are close to doing it—is to charge road users the cost of supplying them with roads, traffic services, and so forth. If this happened in the United States, the electorate would be unlikely to countenance the orgy of road-building that has been sanctioned under the Transport Equity Act, TEA-21, which guaranteed a 47 percent increase in highway funding over five years—much of this for pork barrel projects specially requested by individual congressmen. By charging a proper user fee for roads, the government would raise revenues required to maintain and improve the system.

We need to do all we can to reduce the malign impact of vehicle use: to reduce pollution; to reduce the use of fossil fuels; to reduce congestion; to reduce the damage caused to the environment and to communities who find themselves being torn asunder by roads and ever increasing traffic. Besides insisting that drivers pay the full, direct cost of driving, we should also call on governments to introduce measures that encourage people to use their vehicles more sparingly, and to opt for more fuel-efficient cars. This is a highly complex issue, and a whole range of factors must be taken into account, from improving public transport to imposing heavier taxes on gas-guzzling cars; from introducing road tolls to increasing taxes on gasoline. If we fail to tackle the transport problem, we will end up like a caged hog wallowing in its own excrement.

Large though transport subsidies are, they seem quite modest when compared with subsidies to agriculture, and especially those channeled to farmers in the rich industrialized world. The European Union's (EU's) Common Agricultural Policy (CAP) is a testament to fiscal impropriety, with subsidies amounting to well over US$100 billion a year. And efforts to reduce farm spending in the United States have foundered, with the 2002 Farm Bill sanctioning massive long-term increases in subsidies.

Agricultural subsidies are primarily of benefit to the largest landholders, one third going to just 2 percent of farmers in the United States, and over four-fifths to the top 30 percent. The major winner is corporate agribusiness. Direct payments to farmers mean that the taxpayer foots three-quarters of the agricultural subsidy bill in the United States, whereas in the EU a system of market price supports—involving intervention buying, import taxes, and export subsidies—means that consumers contribute around half of the subsidy.

Abolishing all price supports would be an excellent start to subsidy reform. These supports encourage farmers to produce more food and fiber than the market requires. They also encourage farmers to use excessive inputs of fertilizers and pesticides, and frequently lead to the conversion of marginal land, to soil erosion, and to the loss of wildlife. Furthermore, price supports often result in consumers paying more for their food than they would if they were able to buy produce in an unsubsidized free market. By protecting and favoring domestic producers, governments are also discriminating against producers in other, often poorer, parts of the world. Take sugar, for example. Protectionist policies in the United States and Europe have led to consumers paying higher prices for their sugar, to considerable environmental damage—witness what has happened in the Everglades—and to the impoverishment of sugar producers in the Caribbean, who are not subsidized or protected by their governments.

The 1996 Freedom to Farm Bill attempted to tackle the problem of trade-distorting subsidies, largely by replacing deficiency payments to farmers (these are a form of price support) with direct payments, based initially on past production levels. The intention was to reduce these payments to zero after seven years; but a series of farm crises encouraged the government to abandon its subsidy-slashing plans. Emergency payments and cheap loans meant that the agricultural bill actually rose, rather than fell, and annual bail-outs of over US$7 billion a year were translated into long-term subsidies by the 2002 Farm Bill. By bailing farmers out, year after year, the government is sending out all the wrong signals, and is saying, in effect: you can carry on flooding the market with corn and wheat and oilseed, and we (or, rather, the taxpayer) will help you out when commodity prices plunge, as they inevitably will in times of oversupply. Where is the incentive for farmers to be clever, to try out new crops, to diversify, to produce things that consumers really want, when subsidies such as these exist? There isn't any.

Many of the farmers I met in the United States insisted that all they wanted was a level playing field. Well, the best way to establish a level playing field is to abolish all subsidies. Let us get rid of all the crop programs and export subsidies, the tariffs and low-cost loans, and allow the market to dictate who grows food and where. Bob Buker of the US Sugar Corporation told me when I met him in Florida that it would be unfair if the United States abandoned its sugar program while sugar subsidies remained in place in Europe. I sympathize with his point of view, but believe him to be wrong. Refusing to get rid of your subsidies until other countries abolish theirs is as nonsensical as insisting that you will only provide maternity leave for your working women when other countries promise to do likewise. If we wait for Afghanistan or Burma to adopt such measures, we may have to wait forever.

When I was researching this book, the city of Seattle hosted a meeting of the WTO, which is at the forefront of the move to liberalize world trade, although it is doing so far too tentatively for the likes of the libertarians. There were some spectacular riots in which environmentalists, anti-free traders, anti-child labor groups, trade unionists, and a collection of others took part. There were, undoubtedly, matters to rail against—not least the WTO's obsessive secrecy. But the protests were in many ways inchoate, and an expression of an emotional, gut-felt outrage for the ill-defined phenomenon of globalization, rather than a carefully considered analysis of the issues.

I am, in principle, in favor of trade liberalization. Poor countries need markets, not handouts, and getting rid of trade-distorting subsidies and the protectionist barriers erected by the United States, Europe, and other wealthy countries will generally do them more good than harm. I say generally because the available evidence suggests that trade liberalization brings broad and widely shared economic benefits to those countries who have functioning banks, an independent judiciary, efficient customs services, and the democratic institutions that foster good governance. Trade liberalization is more likely to benefit a small elite, and injure the masses, in countries who are corrupt, undemocratic, and badly governed— and there are plenty of these. But this is not an argument against trade liberalization; it is an argument against rotten government.

Free traders would also argue that the wealthier people become —or, at least, the less poor they become—the more likely they are to take care of their environment. Poverty is often the greatest threat of all to the natural world.

However, I believe that in one important respect the Seattle protesters (or, at least, some of them) were right to argue that free trade is not always fair trade. If you doubt this, then look at the hog industry. The UK's hog farmers are suffering not only because producers elsewhere are subsidized, but because they are having to compete with exports that come from countries where animals have been reared under conditions that are illegal in the UK, and where the costs of production are consequently lower. Even if all subsidies were abolished, the free market would still encourage the worst and cheapest forms of production, and this is unfair on UK producers.

Under the existing rules of the WTO—an organization, incidentally, whose rules are determined by member governments, most of whom are democratically elected and accountable to their voters—countries cannot discriminate between identical products on the basis of how they are produced. One example will suffice to show how the system works. In the mid-1990s, the United States decided to ban the import of shrimp from Thailand, Pakistan, the Philippines, and Malaysia on the grounds that fishermen in those countries used nets that trapped an estimated 20,000 endangered Olive Ridley turtles a year. In 1998, the four countries objected to the US trade ban at a WTO court, and won the case. The WTO ruled that the United States was illegally discriminating against their fishing industries, and it was ordered to lift the import ban.

This may have pleased the Asian fishing communities and their governments; but environmentalists were less than happy. Indeed, they pointed out that the WTO charter actually requires it to promote environmentally responsible trade, and Article 20 suggests that environmental protection can be given priority over market liberalization in incidences where, for example, an endangered species is threatened, as the Olive Ridley turtles were seen to be by fishing practices in Asia. However, the WTO has so far failed to invoke Article 20.

Environmentalists, labor activists, and others—there was even a group called Alien Hand Signals at Seattle—will continue to chafe against the push for sweeping liberalization and to agitate for reforms of the trading system, and quite possibly to riot at future WTO meetings. Meanwhile, many developing countries will continue to maintain that industrial world concerns for the environment, for labor rights, and so forth are a front for protectionism. They fear that taking these issues into consideration will unfairly exclude their participation in the markets that they so desperately need. Inevitably, there is disagreement, too, between various factions within the developed world, and various factions within the developing, with sharp differences of opinion about how far and how fast the process of liberalization should go.

Does this mean that we are destined for endless discord, with the whiff of tear gas and the crash of rocks against riot shields a perennial feature of international debate? Not necessarily. The WTO rules are not written in stone, and can be altered and refined as member governments see fit. I would suggest that one way around the present impasse involves the creation of a two-tier system, with a different set of rules for trade between the 30 industrialized nations who are members of the Organisation for Economic Co-operation and Development (OECD), and trade between OECD countries and the rest of the world. Members of the OECD could invoke environmental, welfare, and labor considerations when it concerned trade between themselves. For example, the UK could insist that hogs destined for the British market should not be reared under systems of production that are illegal in the UK. Likewise, the United States would be able to exclude agricultural produce from other OECD countries if it was grown with pesticides banned in the United States, and it could exclude imports of fish if they were caught using methods that are illegal in the United States. However, when it comes to trade between OECD countries and the rest of the world, then the former would not be able to exclude produce from the latter, or the latter from the former, on environmental or other grounds. Admittedly, such a system would be far from perfect. However, surely it is better to discriminate in favor of the poor, rather than against the poor, and to rely on international laws and conventions, as well as financial transfers from the developed to the developing world, in order to protect dwindling habitats and endangered species in developing countries, to improve labor conditions, and so forth. But whatever trading system is agreed upon, there is no case for retaining subsidies.

Or is there? Many of the environmental commentators who have investigated the subsidy business suggest that there are good subsidies, as well as bad. For instance, David Malin Roodman, the author of *Paying the Piper: Subsidies, Politics and the Environment*, believes that we should "favor broad-gauged incentives and bottom-up approaches to speed the development of environmentally beneficial technologies." In his book *Perverse Subsidies*, Norman Myers likewise suggests that in certain circumstances it may be possible to devise subsidies "to promote the environmental cause," and approvingly cites the agri-environmental payments made by the United States and the EU. These payments encourage farmers to adopt methods of production that protect and enhance the environment and preserve the landscape, and in future they may provide a lifeline for some small farmers who operate in marginal environments, especially in the uplands of Europe. However, Myers concedes that environmental subsidies are problematic, too. For one thing, new subsidies, once in place, are often difficult to remove at a later date when their usefulness has been fulfilled. Subsidies in support of the environment may confuse market choices, like any other subsidies; and—on a pragmatic note—Myers points out that they may well fall foul of existing international regulations.

As it happens, at the time of going to press the EU was debating a plan that involves a significant switch away from production subsidies toward greater support for environmentally friendly farming. The French, who benefit more than any other country from the bloated Common Agricultural Policy, are vigorously opposing the reforms.

This is a highly complex issue, and my own personal feeling is this. Governments should never attempt, through subsidies, protectionist tariffs, or any other means, to determine which industries and which jobs survive. The market should do that, and there are numerous examples—some in this book—to show that dirigist policy-making often has disastrous consequences.

However, there may—I stress may—be a case for providing some agri-environmental grants of the sort envisaged in the latest EU proposals. These would be a payment to farmers for looking after our common heritage; in this case, outstanding landscapes which would be lost if left untended. However, while it is easy enough to promote such an idea in Europe, where the public might be willing to pay farmers modest sums of money to preserve the landscape equivalent of a Rembrandt or a van Gogh, it is hard to see this argument appealing in the United States.

There may also be a case for using fiscal instruments to encourage technological change that is in the best interests of all others. For example, differential taxes could be used to encourage individuals to buy fuel-efficient cars and cleaner fuels. They could also be used to encourage companies to develop renewable sources of energy. But there must always be a caveat: use subsidies as little and rarely as possible, and remember the numerous mistakes previous generations of politicians have made when doling out our hard-earned taxes.

Should subsidy reform be carried out swiftly or gradually? Should industries, farmers, and others who receive subsidies go cold turkey, or spend some time in the financial equivalent of the Betty Ford Clinic, being gently weaned from their addiction? The environmental economist David Pearce points out that shock therapy has been used successfully in countries who are faced by economic crises. Mongolia adopted a rapid program of privatization and market liberalization after the collapse of communism; but Mongolian leaders were fortunate in that the country had no history of political opposition, and the reforms, though hard on many people, went largely unopposed. However, reforming the subsidy system in vibrant democracies such as the United States will be an entirely different matter. True, there are many subsidies that could and should be abolished overnight; but some will be much harder to remove, and Pearce suggests that unless an economic crisis provides the opportunity for wholesale change, subsidy beneficiaries should be given time to adjust. Death by a thousand cuts, as he calls it, is preferable to shock therapy. This will undoubtedly be true of the most subsidized industry of all, agriculture.

One of the immutable truths of global politics is that rich countries do not take kindly to being told to follow the example set by poor countries, so there will be much spluttering on Capitol Hill at the suggestion that political leaders in the United States should take their cue from Tunisia, a small bite-size chunk of the Arab world wedged between Libya and Algeria. But look to Tunisia we should, says Pearce, who admiringly describes the way in which the country reduced food subsidies. Instead of slashing subsidies in one go, which would have hurt the poor and sparked off riots, the government adopted a program of gradual change. It went out of its way to be transparent about the process and forewarned the country of the changes that were going to occur. It also formulated a package of compensatory measures targeted at the most vulnerable groups. One could argue that the Freedom to Farm Bill in the United States attempted to wean farmers off state support in a similar way. However, it failed because the government buckled to farmers' demands for help, and the farmers knew that the government would buckle, so most had little incentive to change the way in which they farmed.

So, can we expect the United States to slash its subsidies? Not nearly as fast as it should. As we have seen, many subsidies are deeply entrenched, and those who receive them will fight tooth and nail to keep them. This brings me back, once again, to the subject of money, and the way in which it is used to buy political favors.

When I first arrived in the United States, I spent time with Anna Aurilio and Lexi Shultz, two women who worked for the US Public Interest Research Group (PIRG), one of many organizations that have sought to highlight the malign influence of money on American politics. Among the publications they gave me was a report, *Subsidies for Sale*, which established the link between corporate contributions and corporate welfare. It listed 289 contributors who had given a total of US$146 million to congressional candidates between 1993 and 1998, "in part to help protect nearly US$41 billion in polluter pork subsidies." There is scarcely a company of any standing that does not figure on the list. For example, beneficiaries of the Army Corps of Engineer's inland waterway system include Archer-Daniels-Midland, which gave US$2,674,414, and Cargill Inc.—big in hogs, as well as corn—which gave US$425,800. General Electric, a major beneficiary of coal-related energy subsidies, contributed over US$2 million and Texaco over US$1 million. Other contributing corporations, all of which benefit from subsidies, included BP, Lockheed, Philip Morris, Bechtel, Kellogg, Monsanto, Chase Manhattan, Goodyear, and Caterpillar. And then there were scores of trade organizations, such as the National Cotton Council, the National Cattlemen's Beef Association, the Michigan Milk Producers Association, and the Georgia Peanut Producers Association; and these, too, were swishing large dollops of money into the coffers of politicians and political parties.

"So the system is corrupt," I suggested to Anna Aurilio when I returned to Washington, DC, toward the end of my travels.

"The system is legal," she said, "so it is not, strictly speaking, corrupt."

One of the more outspoken critics of the campaign contribution system, as it operated at that time, told me that whenever anyone used the word corrupt, the media "turned down the volume"; they were not going to give air time or space to anyone who was so blunt about the issue. And, of course, the media, and especially the television stations, were major beneficiaries: much of the money that politicians received was spent on advertising. However, I would contend that the system was corrupt in the sense that—and for the benefit of libel lawyers, I am using *Chamber's Dictionary*—it tainted, debased, and perverted the democratic system.

That was then; but what about now? Won't the campaign finance reforms piloted through the Senate by Senators Russ Feingold and John McCain, and through the House of Representatives by Chris Shays and Martin Meehan, help to take money-with-strings-attached out of politics—assuming, of course, that court action does not scupper the reforms? Banning soft money and curtailing the use of "issue ads" is undoubtedly a notable achievement, but the limits on hard money contributions have actually been raised, and it was hard money that supplied the bulk of campaign finance last time around. The reforms are welcome, but modest, and those who think that they will break the ties between politicians and big business, or politicians and special interest groups or wealthy individuals, are deluding themselves. The recipients of subsidies—the mining corporations, dam-builders, agribusiness, and timber corporations, to name a few—will continue to find ways of pouring money into the pockets of politicians, and the politicians will continue to vote through programs that make neither environmental nor economic sense.

Politicians frequently argue that they are not corrupted by the campaign contributions they receive—that the decisions they make are in the best interests of the American people. For some this may indeed be true; but for many it quite obviously is not. In the memorable words of Bob Dylan, money doesn't talk, it swears. It ensures that the rich and the powerful get the legislation they need, and it helps them to loot the Treasury, to the detriment of the taxpayer and, frequently, the environment. But as the Republican attorney whom I met in Orlando told me, the people who really deserve our wrath are not the people who get your money; they are the politicians with the backbone of chocolate eclairs who are so easily seduced into perpetuating the subsidy scandal. For as long as politicians and political parties rely on the largesse of corporations and others who can buy political favors with campaign contributions, it is hard to see how the subsidy gravy train can be brought to a halt. I am not suggesting that getting private money out of politics is all we need to do to rid the US, or anywhere else, of perverse subsidies. But it would certainly help.

INDEX